Water, Precious Water

Developed and Published by

AIMS Education Foundation

This book contains materials developed by the AIMS Education Foundation. **AIMS** (**A**ctivities **I**ntegrating **M**athematics and **S**cience) began in 1981 with a grant from the National Science Foundation. The non-profit AIMS Education Foundation publishes hands-on instructional materials that build conceptual understanding. The foundation also sponsors a national program of professional development through which educators may gain expertise in teaching math and science.

- ...is hereby granted ...of it (or the files on ...e used for educational ...accompanying disc may

- ...copy of any portion of ...t of five activities per

- ...copyright information.

A... ...r making more than 200
c... ...n the Internet. Contact us
o...

A...
P.O. Box ...o... ...7 • aimsedu.org

Water, Precious Water contains materials originally developed by the AIMS Education Foundation in cooperation with the California Department of Water Resources and the Irvine Unified School District.

I Hear and
I Forget,

I See and
I Remember,

I Do and

I Understand.
-Chinese Proverb

Water, Precious Water
Table of Contents

Were You Aware?

Topic
Water awareness

Key Question
What are the percentages of salt water, fresh water frozen in glaciers and polar ice caps, and fresh surface and ground water available on Earth?

Learning Goals
Students will:
- predict the percentages of salt and fresh water found on the Earth, and
- compare their predictions to the actual percentages.

Guiding Document
Project 2061 Benchmarks
- *The Earth is mostly rock. Three-fourths of its surface is covered by a relatively thin layer of water (some of it frozen), and the entire planet is surrounded by a relatively thin blanket of air. It is the only body in the solar system that appears able to support life. The other planets have compositions and conditions very different from the Earth's.*
- *Organize information in simple tables and graphs and identify relationships they reveal.*
- *Buttress their statements with facts found in books, articles, and databases, and identify the sources used and expect others to do the same.*

Math
Estimation
Graphing

Science
Earth science
    water

Integrated Processes
Observing
Predicting
Collecting and recording data
Generalizing

Materials
For the class:
    Internet access or reference materials

For each student:
    three colors of colored pencils
    student pages

Background Information
Water is one of the most abundant substances on the face of the Earth, but the amount of fresh water available for use is very limited. Most of the water on the Earth is salt water that is found in the oceans and seas. Salt water makes up over 97 percent of the water on the Earth. Just over two percent of the water on Earth is fresh water frozen in glaciers and the polar ice caps. Less than one percent of the water on Earth is fresh ground water and fresh water in lakes, rivers and streams. About one hundredth of one percent of the water on Earth is in the atmosphere. The majority of this fresh water is not easily obtained, making fresh water a very precious resource. Water is continuously changing form, but the total amount of water on Earth remains constant.

Management
1. Suggested Internet resources are given in the *Internet Connections* section. You may also wish to have a variety of applicable encyclopedias, books, and magazines available for students to use in their research.
2. Students can work individually or in groups to conduct the research.

Procedure
1. As a class, brainstorm as many water sources as possible and categorize them as fresh or salt water.
2. Split the fresh water category into frozen and liquid water categories.
3. Hand out the colored pencils and the first student page. Ask the students to predict the percent of water that would be found in each of three categories; salt water, frozen fresh water in glaciers and the polar ice caps, and fresh surface and ground water.
4. Instruct students to color in their predicted percents on the circular prediction graph. Make sure they start with the type of water they think is most common. For example, if they think salt water is the most common type, and they think it would be about 50 percent, they would need to draw the first line at the 50 mark on the circular prediction graph. Then if they thought fresh water would be 30 percent, they would count 30 marks from the 50 to record the next line. The remaining sector (representing 20 percent) would be the third prediction.
5. Let students share their predictions as well as the reasoning they used to make the predictions.

5

6. Provide time for students to do research on the actual percentages (97% salt water, 2+% frozen fresh water, <1% fresh surface and ground water). Share the results of the research.
7. Distribute the second student page and have them color in the actual percentages in the second graph.
8. Discuss what conclusions can be drawn from the data, and have students write a paragraph on the back of the paper summarizing the discussion.

Connecting Learning
1. What was your prediction for the amount of salt water present on the Earth? …the amount of fresh water that is frozen? …the amount of surface and ground water?
2. How did these predictions compare to the actual values?
3. Were you surprised by the actual values? Why or why not?
4. What do the data tell us? What conclusions can be drawn?
5. If three-fourths of the Earth is covered by water, why is fresh water still such a valuable resource? [Because a vast majority (97%) of the water on Earth is salt water.]
6. What are you wondering now?

Extensions
1. Have students put a cup of slightly salty water in the freezer until it freezes solid. Compare the top and bottom of the ice by tasting both. What do you discover?
2. Look up the local water sources on the Internet or write to a local water agency to find out where fresh water comes from in your area.

Internet Connections
USGS Water Science for Schools
http://ga.water.usgs.gov/edu/
This site has information about the locations of Earth's water, water distribution, the water cycle, rivers, ground water, and more.

Windows to the Universe
http://www.windows.ucar.edu/
Select "Our Planet" and then "Water (The Hydrosphere)" for information about water on Earth. Three reading levels are available—beginner, intermediate, and advanced. It is also available in Spanish.

Enchanted Learning
http://www.enchantedlearning.com/subjects/ocean
This site has information about the oceans with links to other topics like the water cycle and explanations of why oceans are salty.

Were You Aware?

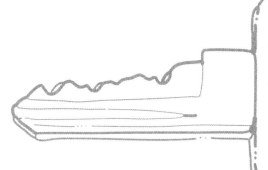

Key Question

What are the percentages of salt water, fresh water frozen in glaciers and polar ice caps, and fresh surface and ground water available on Earth?

Learning Goals

Students will:

- predict the percentages of salt and fresh water found on the Earth, and
- compare their predictions to the actual percentages.

POLAR ICE CAP

Were You Aware?

There are three different types of water available on Earth: salt water, fresh water in glaciers and polar ice caps, and surface and ground water.

Guess the percentages of salt water, fresh water, and frozen water that are on the surface of the Earth. Make a circle graph of your guess. Don't forget to fill in the key!

Key

☐ Salt water

☐ Fresh water

☐ Frozen water

POLAR ICE CAP

Were You Aware?

Graph in the actual percentages of salt water, fresh water, and frozen water that are on the surface of the Earth. On the back of the paper, write a conclusion that you can draw from the data given on the graph.

Key
☐ Salt water
☐ Fresh water
☐ Frozen water

Were You Aware?

Connecting Learning

1. What was your prediction for the amount of salt water present on the Earth? ...the amount of fresh water that is frozen? ...the amount of surface and ground water?

2. How did these predictions compare to the actual values?

3. Were you surprised by the actual values? Why or why not?

4. What do the data you collected tell us? What conclusions can be drawn?

5. If three-fourths of the Earth is covered by water, why is fresh water still such a valuable resource?

6. What are you wondering now?

THE GOODS ON WATER

Topic
Properties of water

Key Question
What properties of water can you observe?

Learning Goal
Students will observe properties of water including its mass and physical states.

Guiding Documents
Project 2061 Benchmarks
- *Measuring instruments can be used to gather accurate information for making scientific comparisons of objects and events and for designing and constructing things that will work properly.*
- *When liquid water disappears, it turns into a gas (vapor) in the air and can reappear as a liquid when cooled, or as a solid if cooled below the freezing point of water. Clouds and fog are made of tiny droplets of water.*
- *Heating and cooling cause changes in the properties of materials. Many kinds of changes occur faster under hotter conditions.*

NRC Standards
- *Objects have many observable properties, including size, weight, shape, color, temperature, and the ability to react with other substances. Those properties can be measured using tools, such as rulers, balances, and thermometers.*
- *Materials can exist in different states—solid, liquid, and gas. Some common materials, such as water, can be changed from one state to another by heating or cooling.*
- *Employ simple equipment and tools to gather data and extend the senses.*

NCTM Standard 2000*
- *Select and apply appropriate standard units and tools to measure length, area, volume, weight, time, temperature, and the size of angles*

Math
Measurement
    estimation
    mass
    volume

Science
Physical science
    matter
        properties of water

Integrated Processes
Observing
Collecting and recording data
Comparing and contrasting
Interpreting data

Materials
For each group:
    3 transparent cups, 9 oz (see *Management 2*)
    2 graduated measuring strips (see *Management 1*)
    tape
    balance
    gram masses
    cold water (see *Management 4*)
    water container (see *Management 3*)
    plastic wrap
    marker

Background Information
Water is one of the most common, yet most distinctive, substances on Earth. Several of its properties can initially be observed using just the senses. Pure water is colorless, odorless, and tasteless, feels wet, and makes recognizable sounds when it is moving, whether a dripping faucet, pelting rain, tumbling waterfall, or crashing waves.

Water, like all matter, has mass and volume. At 4° Celsius, one gram of water has a volume of one milliliter (also called one cubic centimeter or one cc). Therefore, one liter has a mass of 1000 grams (one kilogram).

Whether as an icy solid, a flowing liquid, or an invisible gas (water vapor), water exists in all three physical states under normal conditions on Earth; this is not typical of other substances.

Another of water's interesting properties is that, unlike most liquids, water expands when it freezes. Its mass does not change, but its volume increases nearly nine percent. This means that ice is less dense than liquid water and will float on it. This is why you should leave some air space in a container when freezing a water-based food or beverage.

Management

1. Copy the graduated strips onto transparency film. Each group of four students will need three strips.
2. Each group will need three 9-oz cups. Tape the graduated measuring strips to the outsides of the cups.
3. Recycled water bottles work well for the water containers. Each group will need a bottle with a capacity over 200 mL.
4. The water used in this activity should be cold. Refrigerating the water the night before using it will be adequate.
5. The freezing portion of this activity will take several hours. You may want to complete the activity on the next day.
6. Put the materials for this activity in a central location. To keep the water as cold as possible, do not set the bottles of water out until students need them.

Procedure

Day One

1. Ask students to help you make a list of the properties of water. Tell them that they will be investigating several of the properties that have been listed.
2. Divide the students into groups of four. Show them where the materials are for this activity are located. Distribute the student page.
3. Have students use a marker to label one cup A and another cup B. (The third cup will be used to equalize the balance.) Tell students that it will be important for them to follow the procedures written on the student page.
4. Allow time for students to complete the student page down to the part where they need to freeze one of their cups of water. Inform them that the activity will be completed the following day. Make sure that they cover the cups with plastic wrap to reduce evaporation.

Day Two

1. Before showing the students the cups of frozen water, ask them how they think the cups will compare in mass and in volume. Have a student from each group get a balance and masses.
2. Distribute the cups of frozen water and let students finish gathering their data.
3. Conclude with a discussion about the properties of water.

Connecting Learning

1. What properties of water did you observe with your senses? [colorless, odorless, tasteless, feels wet, takes the shape of its container]
2. What did you discover about the mass of cold water? [1 milliliter equals 1 gram]
3. What would the mass of 50 milliliters of cold water be? [50 grams]
4. If you had 750 grams of cold water, what would its volume be? [750 milliliters] Would this amount fit in your graduated cups? [No.] Explain. [The cups hold about 250 mL.] How many cups would you need? [three]
5. Give examples of places where water is a solid, a liquid, and a gas. [solid—Antarctica, glaciers, icebergs; liquid—lakes, rivers, oceans; gas—water vapor in the air]
6. What are two ways you can use to check if the amounts of water in the two cups are equal? [Place one cup of water on each side of a balance; they should balance. Compare the volumes using a graduated measuring tool; they should have equal volumes.]
7. What did you learn about the mass of a certain volume of water when it is a liquid and when it is frozen? [The mass of the water stays the same whether it is liquid or solid.]
8. What did you learn about the volume of that same water? [The volume of the ice is greater than the volume of the liquid.]
9. Why is it not a good idea to put an unopened can of soda in the freezer over night? [The water in it will expand as it freezes and possibly rupture the can.]
10. In what other situations have you seen something break because the water in it expanded when it froze? [water pipes, water bottles in the freezer, etc.]
11. What are you wondering now?

* Reprinted with permission from *Principles and Standards for School Mathematics*, 2000 by the National Council of Teachers of Mathematics. All rights reserved.

THE GOODS ON WATER

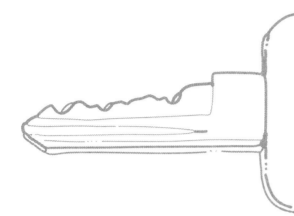

Key Question

What properties of water can you observe?

Learning Goal

observe properties of water including its mass and physical states

Let's see...the frozen water is colder, I know that much.

THE GOODS ON WATER

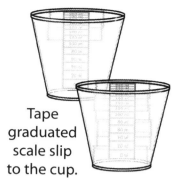

Tape graduated scale slip to the cup.

Graduated Scales
for 9-oz cups

Cover completely with tape.

240	240	240	240	240	240	240	240	240	240
220	220	220	220	220	220	220	220	220	220
200	200	200	200	200	200	200	200	200	200
180	180	180	180	180	180	180	180	180	180
160	160	160	160	160	160	160	160	160	160
140	140	140	140	140	140	140	140	140	140
120	120	120	120	120	120	120	120	120	120
100	100	100	100	100	100	100	100	100	100
80	80	80	80	80	80	80	80	80	80
60	60	60	60	60	60	60	60	60	60
40	40	40	40	40	40	40	40	40	40
20	20	20	20	20	20	20	20	20	20
mL	mL	mL	mL	mL	mL	mL	mL	mL	mL

240	240	240	240	240	240	240	240	240	240
220	220	220	220	220	220	220	220	220	220
200	200	200	200	200	200	200	200	200	200
180	180	180	180	180	180	180	180	180	180
160	160	160	160	160	160	160	160	160	160
140	140	140	140	140	140	140	140	140	140
120	120	120	120	120	120	120	120	120	120
100	100	100	100	100	100	100	100	100	100
80	80	80	80	80	80	80	80	80	80
60	60	60	60	60	60	60	60	60	60
40	40	40	40	40	40	40	40	40	40
20	20	20	20	20	20	20	20	20	20
mL	mL	mL	mL	mL	mL	mL	mL	mL	mL

THE GOODS ON WATER

What properties of water can you observe?

1. Observe water with your senses. Record your group's observations.

2. What is the mass of 100 mL of cold water   Estimate: _____ g
   To find the actual mass:
   a. Place an empty cup on each side of the balance.
   b. Equalize the balance if necessary.
   c. Pour 100 mL of water into one cup.
   d. Find the mass.   Actual: _____ g

   This means that 1 mL of water equals _____ g.

3. What are the three states of water found on Earth?

4. Where on Earth are they found?

5. How does frozen water compare with liquid water?
   a. Fill Cups A and B each with 100 mL of water.
   b. Use the procedure you used in question 2 to find the mass of the water in each cup.

   Cup A _____   Cup B _____

   c. Cover both cups with plastic wrap.
   d. Place Cup A in a freezer for several hours.
   e. When frozen, remove the plastic wrap from both cups.
   f. How do the masses of the two cups compare?

   Cup A _____   Cup B _____

   g. How do the volumes of the two cups compare?

   Cup A _____   Cup B _____

THE GOODS ON WATER

Connecting Learning

1. What properties of water did you observe with your senses?

2. What did you discover about the mass of cold water?

3. What would the mass of 50 milliliters of cold water be?

4. If you had 750 grams of cold water, what would its volume be? Would this amount fit in your graduated cups? Explain. How many cups would you need?

5. Give examples of places where water is a solid, a liquid, and a gas.

6. What are two ways you can use to check if the amounts of water in the two cups are equal?

THE GOODS ON WATER

Connecting Learning

7. What did you learn about the mass of a certain volume of water when it is a liquid and when it is frozen?

8. What did you learn about the volume of that same water?

9. Why is it not a good idea to put an unopened can of soda in the freezer over night?

10. In what other situations have you seen something break because the water in it expanded when it froze?

11. What are you wondering now?

Zooming In

Topic
Water molecules

Key Question
What does a water molecule look like?

Learning Goals
Students will:
- build a model of a water molecule, and
- model hydrogen bonding.

Guiding Documents
Project 2061 Benchmarks
- *Materials may be composed of parts that are too small to be seen without magnification.*
- *All matter is made up of atoms, which are far too small to see directly through a microscope. The atoms of any element are alike but are different from atoms of other elements. Atoms may stick together in well-defined molecules or may be packed together in large arrays. Different arrangements of atoms into groups compose all substances.*
- *Models are often used to think about processes that happen too slowly, too quickly, or on too small a scale to observe directly, or that are too vast to be changed deliberately, or that are potentially dangerous.*

NCTM Standard 2000*
- *Use representations to model and interpret physical, social, and mathematical phenomena*

Science
Physical science
   chemistry
      water molecule

Integrated Processes
Observing
Comparing and contrasting
Making models
Applying

Materials
For each pair of students:
   clay in two colors (see *Management 2*)
   2 or 3 toothpicks
   protractor, optional (see *Management 3*)

Background Information
   The study of molecules of any kind is not easy; it is difficult to comprehend something that is so minute and invisible to the naked eye. Models help because they are tangible; they engage the visual and tactile senses, giving us a better chance at understanding. Models also have drawbacks since they cannot be like the real thing in every detail. Size is one; the model water molecule is much larger than a real water molecule. Lack of motion is another; the water molecule model is static (not moving) while real water molecules are dynamic (in motion). It is important to be aware of the limitations of models so as not to foster naïve conceptions.

Management
1. Students should work together in pairs.
2. Each pair of students will need enough clay to make two one-centimeter balls in one color (representing hydrogen) and one two-centimeter ball in another color (representing oxygen).
3. Protractors are optional. The hydrogen atoms are placed at a 105° angle from the center of the oxygen atom. Students can use a protractor to measure this angle, they can refer to the picture on the student page, or they can use the corner of a paper to find a 90° angle and then make it "just a bit larger."

Procedure
1. Ask the *Key Question* and state the *Learning Goals*. Discuss with students why they may not know what a water molecule looks like. [They're too small to see.]
2. Ask students if the know what water is made up of. [two parts hydrogen and one part oxygen, hence the formula H_2O]
3. Inform students that they are going to use clay and toothpicks to make a model water molecule. Distribute the student page. Inform them of the clay color they should use for the hydrogen and the color for the oxygen.
4. Allow time for them to follow the directions to make the models.
5. Ask them what the water molecule looks like. [a Mickey Mouse head]
6. Together, read the section on hydrogen bonding.

Have groups demonstrate hydrogen bonding by pooling their molecules and connecting several with toothpick halves.

7. Conclude with a discussion about the water molecule and the use of models.

Connecting Learning

1. What are the ingredients of water? [hydrogen and oxygen]

2. How does the size of the hydrogen atom compare to the size of the oxygen atom? [The hydrogen atom is much smaller than the oxygen.]

3. How many hydrogen atoms combine with each oxygen atom to make water? [two]

4. Do you think that all molecules for all substances look like the one for water? Explain. [The structure of molecules for each substance is different. The molecule for salt is structured differently from that of water or of carbon dioxide.]

5. How does the model of a water molecule compare with the real thing? [pluses—it shows three-dimensional structure and relative sizes of the atoms; minuses—the model is far larger than the real molecules and does not show motion]

6. What other models do we use? [the globe, cells, water cycle, planets, phases of the moon, etc.]

7. Why do we use models? [Models are used to help us see how things that are too big or too small are structured or how they work.]

8. What are you wondering now?

* Reprinted with permission from *Principles and Standards for School Mathematics*, 2000 by the National Council of Teachers of Mathematics. All rights reserved.

Zooming In

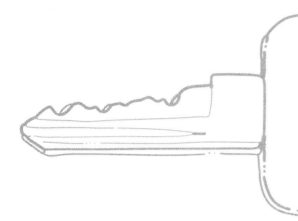

Key Question

What does a water molecule look like?

Learning Goals

Students will:

- build a model of a water molecule, and
- model hydrogen bonding.

Zooming In

Water, like everything else, is made of atoms. Atoms join together in different combinations to form molecules. Water molecules have two hydrogen atoms and one oxygen atom. That's why water is called H_2O.

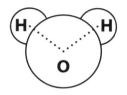

Build a model of a water molecule.

(H)   For hydrogen, make two 1-cm clay balls in one color.

(O)   For oxygen, make one 2-cm clay ball in another color.

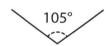

Press the two hydrogen atoms onto one oxygen atom so they from an angle of about 105°.

Water molecules are attracted to each other. The hydrogen atoms of one molecule are attracted to the oxygen atoms of other molecules. This attraction or pull toward each other is called hydrogen bonding. The water molecules are on the move, constantly breaking and reforming hydrogen bonds.

Gather several models of water molecules. Show hydrogen bonding using toothpicks broken in half.

hydrogen bonds

How does this model compare with a real water molecule?

22

Connecting Learning

1. What are the ingredients of water?

2. How does the size of the hydrogen atom compare to the size of the oxygen atom?

3. How many hydrogen atoms combine with each oxygen atom to make water?

4. Do you think that all molecules for all substances look like the one for water? Explain.

5. How does the model of a water molecule compare with the real thing?

6. What other models do we use?

7. Why do we use models?

8. What are you wondering now?

Topic
Properties of water

Key Questions
1. What evidence do you observe that water is attracted to itself or something else?
2. How do the surface tensions of water and soapy water compare?

Learning Goals
Students will:
- do a variety of simple investigations to observe the attraction of water, and
- read about other distinctive properties of water.

Guiding Documents
Project 2061 Benchmarks
- *Scientific investigations may take many different forms, including observing what things are like or what is happening somewhere, collecting specimens for analysis, and doing experiments. Investigations can focus on physical, biological, and social questions.*
- *Materials may be composed of parts that are too small to be seen without magnification.*

NRC Standards
- *Scientists develop explanations using observations (evidence) and what they already know about the world (scientific knowledge). Good explanations are based on evidence from investigations.*
- *Employ simple equipment and tools to gather data and extend the senses.*
- *Earth materials are solid rocks and soils, water, and the gases of the atmosphere. The varied materials have different physical and chemical properties, which make them useful in different ways, for example, as building materials, as sources of fuel, or for growing the plants we use as food. Earth materials provide many of the resources that humans use.*

Science
Physical science
  matter
    properties of water
      cohesion
      adhesion

Integrated Processes
Observing
Collecting and recording data
Comparing and contrasting
Interpreting data

Materials
For each group:
  copy paper strips in two colors, 2 cm x 18 cm
  two transparent cups, 9 oz
  ruler
  tape
  two paper clips
  small plastic plate
  wax paper
  eyedropper
  toothpick
  two transparent straws
  scissors
  soft paper towel
  string, 1 m

For the class:
  liquid dish soap
  containers for water (see *Management 3*)

For each student:
  Water, Water Everywhere rubber band book
  #19 rubber band
  student pages

Background Information
Water has several distinctive properties that are important in how our world functions. In this activity, students will observe cohesion and adhesion, and other properties will be introduced through a booklet they assemble.

Cohesion is the attraction between like molecules, in this case, water molecules to water molecules. Water molecules are composed of two hydrogen atoms and one oxygen atom. The hydrogen atoms are attracted to the oxygen atoms of other water molecules (hydrogen bonding).

At the water's surface, the molecules have little attraction to the more distant molecules of water vapor in the air, so the pull of molecules to the sides and downward dominates. Molecules are continually being pulled down and replaced by others, maintaining the minimum surface area possible. This *surface tension*—the pull of surface molecules toward other water molecules—acts like a kind of skin on which a paper clip can float. It causes a drop of water to form a nearly spherical shape and water to develop a dome when a glass is filled past the rim. Surface tension is the evidence of the cohesiveness of water.

Water has high surface tension. When soap is introduced, the soap molecules move between the water molecules, weakening their hydrogen bonds and reducing surface tension by two-thirds. The lower surface tension makes the skin more elastic, so soapy water drops are flatter than water drops and soap bubbles do not break apart as quickly as water bubbles.

Water is also adhesive; it clings to other substances like glass, soil, paper, and plant and animal cells. *Adhesion* is the attraction between unlike molecules. The adhesive force overcomes the force of gravity when water climbs up a paper towel strip or climbs the surface of a straw, forming a concave curve, or meniscus. Because of adhesion, water drops cling to a cup, to string, and to a toothpick.

Management
1. Divide the class into groups of two or three.
2. Put the materials on a table so that groups can come and gather what they need.
3. Use pitchers or clean two-liter bottles as water containers. Fill the containers with tap water. Add a few drops of liquid dish soap to one container and label it "soapy water."

Procedure
Part One
1. Inform the class that water has some unusual properties. They will be looking for proof that water is attracted to itself. Distribute the first page of *Part One* and have groups gather the necessary materials.
2. Guide groups in constructing the *Water Snakes*. Instruct them to follow the directions and record their observations. Ask what this shows about water.
3. Have students float the paper clip and overfill the cup, record their observations, and describe what these activities show about water. Explain that surface tension—the pull of surface molecules toward other water molecules—makes the water act as if it has a skin. It allows the paper clip to float, even though it is more dense than water. Surface tension also lets us fill a cup with water above its rim.
4. Ask, "How do the surface tensions of water and soapy water compare?" Field predictions, then give students the second page of *Part One*. Have them gather materials, follow the directions, and discuss the results.

Part Two
1. Point out that, so far, students have observed that water molecules are attracted to other water molecules. This attraction of like molecules is called cohesion. It is what causes surface tension. Now they will look for another kind of attraction. Distribute the first page of *Part Two*.

2. Have students gather materials and do the three activities with the cup, waxed paper, and straw. Ask what these observations show about water.
3. Distribute the second page of *Part Two*. Have students gather materials and carry out the paper towel and string investigations. Ask what these observations show about water.
4. Explain that water is not only attracted to itself; it is attracted to other molecules like glass, string, and paper. The attraction of unlike molecules is called adhesion. Was water attracted to the waxed paper? [No, it is not attracted to everything.]
5. Give students the *Water, Water, Everywhere* rubber band book and have them assemble it.
6. In pairs, small groups, or as a class, read the rubber band book to review cohesion and adhesion and learn about other properties of water.

Connecting Learning
1. What do the observations in *Part One* show about water? [Water has cohesion—the attraction between like molecules.] How do you know? [The water snake is attracted to the water, but the dry snake isn't. The paper clip floats because the water acts like it has a skin on the surface. The surface skin keeps water from spilling over the sides of the cup.]
2. Did all the groups have the same experience with the paper clip? If not, why do you think your experience was different? (The paper clip may not float if the wire does not form a closed curve.)
3. Which has more surface tension—water or soapy water? [Water has higher surface tension than soapy water.] How do you know? [Soapy water drops are flatter and more spread out than plain water drops and soap bubbles last longer than water bubbles because soap weakens the amount of pull.]
4. What do the observations in *Part Two* show about water? [Water has adhesion—the attraction between unlike molecules. It is attracted to some, but not all, materials.] How do you know? [Water drops cling to the plastic cup. Water drops are attracted to a toothpick and can be led across wax paper. Water climbs up the straw's surface and the paper towel, overcoming the force of gravity. During the string pour, water clings to the string (adhesion); water molecules also cling to each other (cohesion), reducing the possibility they will drop off the string.]
5. Do you think you could position the string so that the force of gravity would be stronger than the adhesive force? Explain. [If the string is held parallel to the floor, the water will not make it the entire length of the string before falling.]
6. Which investigations were most interesting to you? Which one would you like to show your family?
7. What are you wondering now?

Key Questions

1. What evidence do you observe that water is attracted to itself or something else?
2. How do the surface tensions of water and soapy water compare?

Learning Goals

Students will:

- do a variety of simple investigations to observe the attraction of water, and

- read about other distinctive properties of water.

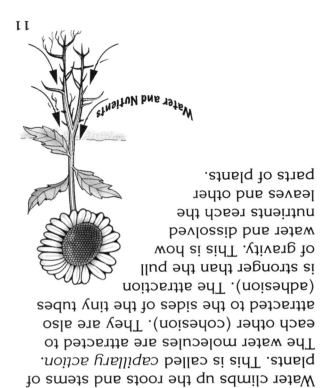

Water and Nutrients

Water climbs up the roots and stems of plants. This is called *capillary action*. The water molecules are attracted to each other (cohesion). They are also attracted to the sides of the tiny tubes (adhesion). The attraction is stronger than the pull of gravity. This is how water and dissolved nutrients reach the leaves and other parts of plants.

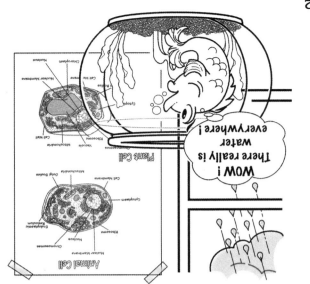

WOW! There really is water everywhere!

Water is one of the most common substances on Earth. It is also very special. It is in the air, on the surface, and underground. Water is also a large part of the cells of most plants, animals, and humans. Let's take a look at some of the unusual properties of water.

Water is very precious. Living things need to take in water to survive and grow. For some, like fish or sphagnum moss, water is also their habitat. Without water, there would be no life on Earth.

Water, Water, Everywhere

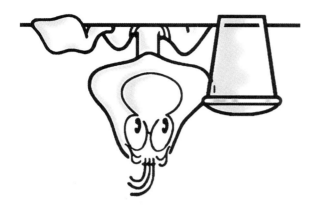

Because water has high surface tension,
- water forms nearly round drops,
- paper clips can float on its skin,
- a cup can be filled above the rim, and
- some insects can skate across its surface.

Water expands and becomes less dense when it freezes. (Most other liquids contract when frozen.) That's why ice floats. That's also why ice forms on the surface of a lake rather than on the bottom. Surface ice reduces the amount of heat that escapes from the water underneath. This helps plants and animals that live there survive.

Water is also attracted to other substances like glass, plant and animal cells, cloth, and soil. This property is called *adhesion*. Look at how water clings to the surface of a glass, in spite of gravity. Notice how water climbs up a paper towel.

Water can be a gas, a liquid, or a solid in normal temperatures on Earth. It is the only natural substance for which this is true. During a storm, there can be water vapor, rain, and hail in the air at the same time.

29

A water molecule has two hydrogen atoms and one oxygen atom. Its scientific name is H_2O. The hydrogen atoms are attracted to the oxygen atoms of other water molecules. The pull of like molecules toward each other is called *cohesion*. The molecules are in motion, so bonds are constantly being made and broken.

Water takes in and gives off large amounts of heat energy slowly. Ocean temperatures have been as cold as -2°C and as warm as 36°C. Land temperatures have been as cold as -88°C and as hot as 58°C.

Color the temperature ranges on the graph. You can see that ocean temperatures change less than temperatures change on land. Oceans help control the temperature on Earth.

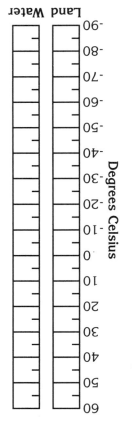

Above the surface, water molecules are too far away to exert much pull. Under the surface, water molecules are pulling toward each other from every direction. At the surface, water molecules are pulling to the sides and down. The tension at the surface makes water act as if it is covered with a film or skin.

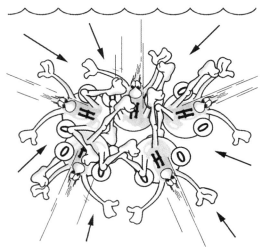

Water dissolves more substances that any other liquid. It is sometimes called the "universal solvent." Sugar dissolves in the water we use to make lemonade. Salt dissolves in the water in soup. Nutrients from the foods we eat dissolve in the water in our cells. Minerals dissolve in the water as it travels through rocks and soil.

Does sugar dissolve faster in hot water or in cold?

What do these observations show about water?

Materials
Paper strips in two colors
Scissors
Transparent cup with water
Ruler
Tape
Two paper clips
Small plastic plate
Liquid dish soap

Water Snakes

- Cut and fold the paper strips like an accordion. Firmly crease the folds. Cut a snake's head on one end of each strip. Tape the snakes to opposite ends of the ruler.

- Nearly fill the cup with water. Dip one snake's head about ½ cm into the water.

- Slowly lower the dry snake toward the water until it is barely above the water's surface. Repeat with the wet snake.

What do you observe?

Floating Paper Clips

- Drop one paper clip into the water. Observe. Remove and dry the paper clip.

- Bend the second paper clip as shown and use it to carefully lower the unbent clip onto the surface. Observe.

- Add a drop of soap. Observe.

- Try the same thing with soapy water. Observe.

How did your observations compare?

Overflowing Cup

- Put the cup in a plate

- Fill the cup to the top with water.

- Add more water a little at a time, without letting the cup overflow.

What do you observe?

Part One B
Cohesion

How do the surface tensions of water and soapy water compare?

Materials
Wax paper
Eyedropper
Toothpick
Cup with water
Cup with soapy water
2 straws

Put a drop of water onto wax paper.

Draw the side view. ⟶ _____
water

Dip the tip of a toothpick into soap and touch it to the water drop

Draw the side view. ⟶ _____
soapy water

What do you observe?

Put a straw in the cup of water and blow. What do you observe?

Put another straw in the soapy water and blow. What do you observe?

Water	Soapy water

WATER, PRECIOUS WATER   32   © 2009 AIMS Education Foundation

What do these observations show about water?

Pour water into a transparent cup.
Then pour it out and look at the cup.

What do you observe?

Materials
Transparent cup
Wax paper
Eyedropper
Toothpick
Transparent straw
Water

Use an eyedropper to put drops of water on wax paper.
Try to move the water drops with a toothpick.

What do you observe?

Dip a straw about 2 cm into the water.
Seal the top with a finger and lift it up,
but not out of the water.

What do you observe about the shape
of the water in the straw?

Draw what
you observe.

What do these observations show about water?

Cut the paper towel the size and shape of the pattern. Dip the point just barely into the water and keep it there.

What do you observe?

Materials
2 transparent cups
Scissors
Soft paper towel
1 meter of string
Tape
Water

Outdoors, hold a cup of water about one meter above the ground and over the other cup. Slowly pour the water into the empty cup.

Return the water to the upper cup. Wet the string and tape one end inside each cup. Have your partner move the lower cup about a meter to the left and pull the string taut. Slowly pour water from the upper to the lower cup.

What do you observe?

On the picture, draw where the water goes.

34

Connecting Learning

1. What do the observations in *Part One* show about water? How do you know?

2. Did all the groups have the same experience with the paper clip? If not, why do you think your experience was different?

3. Which has more surface tension—water or soapy water? How do you know?

4. What do the observations in *Part Two* show about water? How do you know?

35

Connecting Learning

5. Do you think you could position the string so that the force of gravity would be stronger than the adhesive force? Explain.

6. Which investigations were most interesting to you? Which one would you like to show your family?

7. What are you wondering now?

Bubble Busters

Topic
Properties of water

Key Question
How does the amount of soap in a solution affect the time a bubble will last?

Learning Goal
Students will use scientific processes to investigate the time bubbles made with different amounts of soap last.

Guiding Documents
Project 2061 Benchmarks
- *Tables and graphs can show how values of one quantity are related to values of another.*
- *Find the mean and median of a set of data.*
- *Recognize when comparisons might not be fair because some conditions are not kept the same.*

NRC Standards
- *Design and conduct a scientific investigation.*
- *Use mathematics in all aspects of scientific inquiry.*

NCTM Standards 2000*
- *Select and apply appropriate standard units and tools to measure length, area, volume, weight, time, temperature, and the size of angles*
- *Carry out simple unit conversions, such as from centimeters to meters, within a system of measurement*
- *Represent data using tables and graphs such as line plots, bar graphs, and line graphs*
- *Use measures of center, focusing on the median, and understand what each does and does not indicate about the data set*

Math
Measurement
    elapsed time
Data analysis
    measures of center (median or mean)
Graphing
    bar

Science
Scientific inquiry
Physical science
matter
    properties of water
        surface tension

Integrated Processes
Observing
Predicting
Identifying and controlling variables
Collecting and recording data
Drawing conclusions

Materials
For the class:
    four containers, each with 1 liter of water
    bottle of quality liquid dish soap
    measuring spoons (see *Management 1*)
    water
    liter measuring tool

For each group:
    four straws
    plastic plates, 6-inch (see *Management 3*)
    four cups, about 5 oz
    stopwatch or watch with second hand
    paper towels

Background Information
Surface tension
    The surface of water acts like a tight skin; it pulls together or contracts into the smallest surface area possible. This *surface tension* exists because water molecules are attracted to each other; they are cohesive. Below the surface, they are attracted equally in all directions. At the surface, they are attracted to the sides and downward.
    The surface tension of water is greater than that of other common liquids. Try blowing a bubble with water. It breaks apart because the surface tension is so strong. Soap weakens the bonds between water molecules, reducing the surface tension to about one-third that of plain water. This makes the soapy water more elastic; it stretches to form bubbles when air is blown into it.

Variables
    Both gravity and evaporation affect the life of a bubble. Gravity causes soap solution to drain down the sides, leaving a thinning area on top. Evaporation thins the entire surface of the bubble, causing it to eventually burst. The rate of evaporation increases when the air is dry (low humidity), temperatures are warm, and air currents from wind or air-conditioners are blowing. Vibrations disturb the stability of the soap film, making it more likely to burst. Additional variables include the soap solution—kind of soap, amount of soap, additives—as well as bubble size.

Management

1. If metric measuring spoons are unavailable, substitute the number of teaspoons listed in *Management 2*. One teaspoon is equal to 5 mL.
2. Label four containers A, B, C, and D and fill each with one liter of water. Put 3 mL (½ teaspoon) of liquid dishwashing soap in A, 25 mL (5 teaspoons) in B, 75 mL (15 teaspoons) in C, and 150 mL (30 teaspoons) in D.
3. To control bubble size, groups will blow bubbles in six-inch plastic plates. Each group needs four plates—one for each solution. As soon as the bubble has expanded to fill the entire plate, students can stop blowing.
4. Divide students into groups of four; each member can blow three of the 12 bubbles. Each group will need four cups, labeled A, B, C, and D, filled with the four soap solutions.
5. Lay materials out for students to gather.
6. Before blowing bubbles, wet the plates with soap solution. A dry surface will cause the bubbles to burst.

Procedure

1. Ask, "How long do you think a bubble might last? (Take estimates.) What might affect how long it lasts?" (List their ideas.) Explain that the variable they will be testing is the amount of soap. Inform the class that soap reduces the surface tension of water so that the bubble skin can stretch more and not pop right away.
2. Tell students they will blow bubbles on a plate and time how long they last. Ask, "From our list, what variables should we control?" [air currents, vibrations, kind of soap, bubble size]
3. To control bubble size, show students how to blow a bubble on a plastic plate. Demonstrate how to pour a small amount of bubble solution on the plate to coat it, place the straw just above the surface at an angle, and gently blow until the bubble fills the indented portion of the plate. Have someone time how long it takes for the bubble to burst. Discuss how to record elapsed time in seconds (2 minutes, 36 seconds = 156 seconds).
4. Have each group gather four straws, the four cups of soap solution, four plastic plates, a stopwatch, and several paper towels. Distribute the first activity page.
5. Call attention to the amount of soap in each solution. Ask students to predict the order of the solutions (A, B, C, D) from longest-lasting to shortest-lasting bubbles, and write why they chose this order. Have them also record the variable tested and those controlled.
6. Give groups the table and graph page. Point out that they will blow and time three bubbles for each solution. Have them decide on a plan to equitably involve all group members in timing and blowing bubbles.

7. After the data are gathered, have students record either the median or calculate the mean. Ask them to look at the range of the data, decide on the graph increments (5, 10, 20, or 25 seconds), and complete the bar graph.
8. Return to the first page and have students record the actual results. Discuss their observations and conclusions.
9. To gather a larger, more reliable data sample, have all of the groups report their original times for each solution and determine the median and/or mean of the pooled data. Compare these averages with individual group averages and discuss how smaller samples can be problematic.

Connecting Learning

1. How would you describe the bubble skin? [thin, transparent, elastic, colorful, etc.]
2. What is inside the bubble? [air] Why is it important for the bubble? [Air is pushing the bubble skin outward. Without it, there would be no bubble.]
3. What shape does the bubble form on a flat surface? [hemisphere]
4. What changes did you observe in the bubbles? [The top got thinner. Colors changed until some black spots appeared at the top and it popped. (In high school physics, students will learn that the color changes are due to interference—how light waves behave as they reflect off the surfaces of thinning soap film.)]
5. What was our manipulated variable? [amount of soap] What variables did we control? [kind of soap, size of bubbles, air currents, vibrations]
6. What were your results? Were the results expected or surprising? Explain.
7. Adding more soap always produces longer-lasting bubbles. Agree or disagree? Explain.
8. How did the class averages from pooled data compare with the averages in individual groups? What might cause these differences?
9. What are you wondering now?

Extensions

1. Suggest students investigate whether different kinds of dish soap affect how long a bubble lasts.
2. Experiment with adding glycerin or corn syrup to the soap solution.
3. Have a friendly contest to see who can blow the largest bubble.
4. Based on the results of this activity, have the class devise a plan for finding the optimum amount of soap for producing the longest-lasting bubbles. Assist them in carrying out their plan, organizing the data, and drawing conclusions.

* Reprinted with permission from *Principles and Standards for School Mathematics,* 2000 by the National Council of Teachers of Mathematics. All rights reserved.

Key Question

How does the amount of soap in a solution affect the time a bubble will last?

Learning Goal

use scientific processes to investigate the time bubbles made with different amounts of soap last.

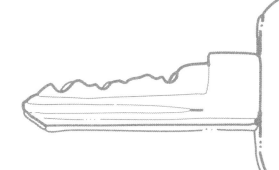

How does the amount of soap affect the time a bubble will last?

Predict Rank the solutions in order from longest-lasting to shortest-lasting bubbles. Why did you rank them this way?

Prediction _____   _____   _____   _____

longest-lasting   shortest-lasting

Actual _____   _____   _____   _____

Control variables What variable are we testing?

What other variables could affect this investigation?

Collect data Do three time trials for each solution, determine either the median or mean average, and complete the bar graph.

Interpret data Above, record the actual order of the solutions. What are your conclusions?

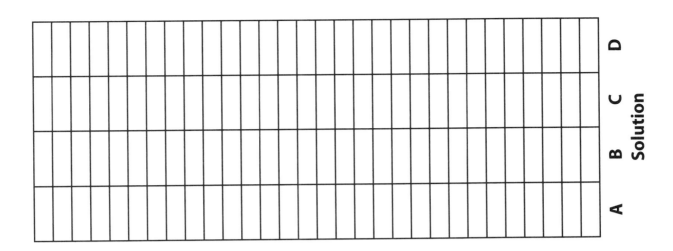

BUBBLE BUSTERS

Time in Seconds

A   B   C   D
Solution

Solution one liter water	A 3 mL soap	B 25 mL soap	C 75 mL soap	D 150 mL soap
Trial 1 time in seconds				
Trial 2 time in seconds				
Trial 3 time in seconds				
Total Time				
* _____ **Average Time**				

* Median or Mean

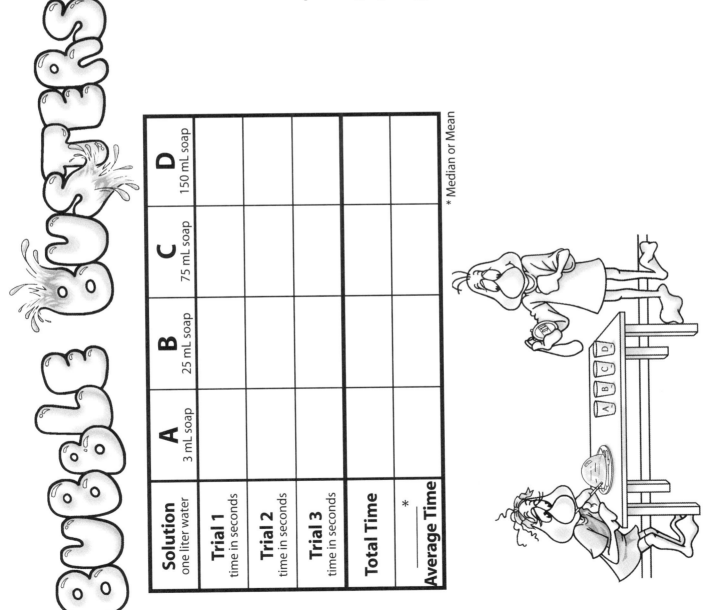

Connecting Learning

1. How would you describe the bubble skin?

2. What is inside the bubble? Why is it important for the bubble?

3. What shape does the bubble form on a flat surface?

4. What changes did you observe in the bubbles?

5. What was our manipulated variable? What variables did we control?

Connecting Learning

6. What were your results? Were the results expected or surprising? Explain.

7. Adding more soap always produces longer-lasting bubbles. Agree or disagree? Explain.

8. How did the class averages from pooled data compare with the averages in individual groups? What might cause these differences?

9. What are you wondering now?

Measure Up

Topic
Volume measurement

Key Question
How many centimeters high will the 100-milliliter mark be on your science cup?

Learning Goals
Students will:
- calibrate a cup,
- observe conservation of volume, and
- use the calibrated cup as a measuring tool.

Guiding Documents
Project 2061 Benchmarks
- *Measuring instruments can be used to gather accurate information for making scientific comparisons of objects and events and for designing and constructing things that will work properly.*
- *Measure and mix dry and liquid materials (in the kitchen, garage, or laboratory) in prescribed amounts, exercising reasonable safety.*

NRC Standard
- *Employ simple equipment and tools to gather data and extend the senses.*

NTCM Standards 2000*
- *Understand that measurements are approximations and understand how differences in units affect precision*
- *Select and apply appropriate standard units and tools to measure length, area, volume, weight, time, temperature, and the size of angles*

Math
Measurement
    height (cm)
    volume/capacity (mL)

Integrated Processes
Observing
Predicting
Collecting and recording data
Comparing and contrasting

Materials
For each student:
    one transparent cup, 10 oz
    paper clip
    student page

For the class:
    water
    liquid dish soap
    corn syrup
    cubic-centimeter blocks (see *Management 1*)

For each group:
    graduated cylinder (see *Management 2*)
    several strips of masking tape
    eyedropper (see *Management 3*)
    water container such as a 16-oz cup
    mystery cup (see *Management 4*)

Background Information
To calibrate a cup means to measure and mark the graduations. In this activity, students will measure 20 milliliters of water, pour it into a cup, and mark the water level, repeating this process until the scale reaches the top of the cup. The cup can then be used as a measuring tool.

Measurement is approximate. Any measuring unit can be divided into smaller, more precise units; a liter, for example, can be divided into milliliters (thousandths) or even smaller units called microliters (millionths). How carefully a person measures the water and marks the water level each time also affects preciseness.

Liquid takes the shape of its container, so the container's size and shape determine the height of a given volume of liquid. The same volume of water falls at different heights in different containers. *Conservation of volume* means the volume remains the same even when the shapes of the containers are different.

In this case, volume is the amount of space water occupies. Capacity is the amount a container can hold. The tallest or widest container does not necessarily have the greatest capacity.

Management
1. To help students visualize a milliliter, gather blocks that are one cubic centimeter—centicubes or unit blocks from a base ten set. One milliliter of water equals one cubic centimeter.
2. Students will need graduated cylinders to measure 20 mL of water. Do not use graduated cylinders larger than 100 mL. Smaller cylinders will produce more accurate results.
3. For more precise measurements, use an eyedropper to add or remove water a little under or over the 20-mL line. Read the measurement at eye level.

4. The mystery cups can be any clear plastic cups that are shaped differently from the science cups. Nine-ounce cups with slanted sides are suggested.
5. Put out the materials for students to collect. Groups will need a water source to use when filling their beakers to calibrate the cups.
6. Divide the class into groups of three. Students will individually calibrate their science cups and, as a team, calibrate the mystery cup.
7. To measure height, use your own rulers or the ruler on the page. Round to the nearest millimeter and record as a decimal.

Procedure

1. Hold up a centicube. Explain that one milliliter has the same volume as a cubic centimeter. Gather several centicubes so students can visualize several milliliters. Display a 10-oz cup and ask, "If I pour 100 mL of water in this cup, how many centimeters high will the water level be?"
2. Give students the activity page and have them record their predictions—both 100-mL height and capacity—for the science cup.
3. Announce that each student will calibrate his or her own 10-oz cup. Have each group gather water, masking tape, an eyedropper, 10-oz cups, and a 20-mL graduated cylinder. Instruct students to place strips of masking tape on the outsides of their cups from top to bottom.
4. Have group members take turns measuring 20 mL of water, pouring it into their respective cups, and marking and labeling the water level. Tell them to continue marking the 20-mL measures until their cups are calibrated.
5. Direct students to measure the height of the 100-mL mark and record both the height and the cup capacity. Ask them to compute the difference between predictions and actual measurements. Save the cup for future use.
6. Distribute a mystery cup to each group. Have them record their predictions of the 100-mL height and cup capacity.
7. Without teacher input, ask groups to calibrate the mystery cup, record the 100-mL height as well as cup capacity, and compute the differences between predictions and measurements.
8. Have each group member use the science cup to measure one ingredient in the bubble recipe. The solution should total 120 mL.
9. Invite students to form bubble makers from paper clips or wire and blow bubbles.

Connecting Learning

1. How can we prove our measurements are approximate? [Compare the scale marks on several 10-oz cups.] Try it. Why does this happen? [It depends on how precisely each person measures the water and marks the water level. Even a careful person will have an approximate measurement, though, because any measurement unit can be divided into smaller, more precise units.]
2. Compare 100 mL of water in the science cup and the mystery cup. What is the same? [volume] What is different? [the height of the 100-mL mark] Why? [Water takes the shape of its container, so the size and shape of the cup—in this case, the diameter and wall angle—determine the height of a given volume of water.]
3. How does the shape of the cup affect the 100-mL level? [The narrower the glass, the higher the 100-mL level will be.]
4. How do the capacities of the science cup and the mystery cup compare?
5. For what other things could we use the calibrated cup? (See *Extension 1* for ideas.)
6. What are you wondering now?

Extensions

1. Use the calibrated cups to:
   • make gelatin and "finger gelatin." Note the amount of water in the two recipes. Discuss how water conservation can begin with something as simple as gelatin.
   • follow any recipe requiring small volume measurements.
   • mix paints (dry tempera and water).
2. Calibrate a 10-oz cup using a balance and gram masses. One milliliter of cold water has one gram of mass. Place the empty cup on one side of the balance and equalize with gram masses on the other side. This adjusts for the mass of the empty cup. Then add 20 grams. Pour water into the cup until both sides balance. Mark the level. Continue adding 20 grams at a time until the cup is calibrated.

* Reprinted with permission from *Principles and Standards for School Mathematics*, 2000 by the National Council of Teachers of Mathematics. All rights reserved.

Measure Up

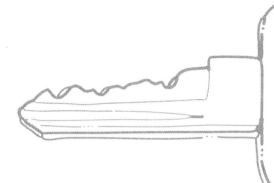

Key Question

How many centimeters high will the 100-milliliter mark be on your science cup?

Learning Goals

Students will:

- calibrate a cup,
- observe conservation of volume, and
- use the calibrated cup as a measuring tool.

Measure Up

The Science Cup

	Height at 100-mL	Capacity of cup
Prediction	cm	mL
Actual	cm	mL
Difference	cm	mL

The Mystery Cup

	Height at 100-mL	Capacity of cup
Prediction	cm	mL
Actual	cm	mL
Difference	cm	mL

BUBBLES Recipe

Make your own bubble solution.

20 mL liquid dish soap
20 mL corn syrup
80 mL water

Have fun blowing bubbles!

Measure Up

Connecting Learning

1. How can we prove our measurements are approximate? Try it. Why does this happen?

2. Compare 100 mL of water in the science cup and the mystery cup. What is the same? What is different? Why?

3. How does the shape of the cup affect the 100-mL level?

4. How do the capacities of the science cup and the mystery cup compare?

5. For what other things could we use the calibrated cup?

6. What are you wondering now?

Topic
Volume measurement

Key Question
Which bottle do you think holds the most water?

Learning Goal
Students will compare the capacities of four different bottles, predicting and observing the various volumes of water each can hold.

Guiding Documents
Project 2061 Benchmarks
- *Tables and graphs can show how values of one quantity are related to values of another.*
- *Organize information in simple tables and graphs and identify relationships they reveal.*

NCTM Standards 2000*
- *Understand such attributes as length, area, weight, volume, and size of angle and select the appropriate type of unit for measuring each attribute*
- *Select and apply appropriate standard units and tools to measure length, area, volume, weight, time, temperature, and the size of angles*
- *Collect data using observations, surveys, and experiments*
- *Represent data using tables and graphs such as line plots, bar graphs, and line graphs*

Math
Measurement
  volume/capacity (mL)
Graphing

Integrated Processes
Observing
Predicting
Comparing and contrasting
Collecting and recording data

Materials
For each group:
  four bottles, each a different shape
  pitcher of water
  graduated cylinder
  paper towels

Background Information
The capacity of a container is the amount it can hold. This is measured in milliliters (mL). Estimating the capacity of irregularly-shaped containers is not easy. While it seems instinctive to think in terms of size, it is not always true that the tallest or widest bottle holds the greatest volume. In this activity, students will develop their concept of conservation of volume as they make predictions about the capacities of various bottles and then test their predictions.

Management
1. Students should work together in groups of four to six.
2. Before doing this activity, you will need to collect a variety of empty containers of different sizes and shapes. Ask students to help you with the collection by bringing from home items such as soda bottles, cottage cheese containers, shampoo bottles, canning jars, or syrup bottles.
3. Group bottles of similar size but different capacities into sets of four. Use tape or stickers to label the sets of bottles A, B, C, and D. Cover any markings that indicate capacity.
4. Each group will need a graduated cylinder or a similar container that will allow them to measure the capacity of each of their bottles in milliliters.
5. You can add a problem-solving component to this activity by including bottles that hold more than the graduated cylinders.

Procedure
1. Display four bottles of similar size and ask, "Which bottle do you think holds the most water? (Listen to responses.) How could we find out?" [measure the amount of water they hold]
2. Divide students into their groups and distribute the materials and the first student page.
3. Have students sketch their groups' containers in the appropriate spaces and predict which container will hold the most and which will hold the least.
4. Instruct each group to fill the first bottle with water from the pitcher and then pour the water from the bottle into the graduated cylinder to determine the volume.
5. Have groups repeat this process with each of their bottles, recording the data on their student pages.

6. Once all data are collected, have students record the actual order that the bottles go in from greatest to least capacity.
7. Hand out the second student page, and have students complete the graph using the data they just collected.
8. Discuss the differences between students' predictions and the actual results.
9. Have each group bring the bottle that held the most water to the front of the class.
10. Re-label these bottles a, b, c, d, e... and distribute the third student page.
11. Have students predict the order of the bottles by capacity, record the capacity reported by each group, sketch the *Super Bottle* with the greatest capacity, and order the rest of the bottles by actual data.
12. Discuss the differences between predictions and the actual results.

Connecting Learning

1. What property of water did you observe? [It takes the shape of its container.] What other matter has this same property? [all kinds of liquids]
2. How did your predictions compare with the actual results?
3. Did your predictions improve as you had more experience? Why or why not?
4. What do you notice about the bottle that holds the most water? ...the least water?
5. How does shape affect capacity? [Tall and wide holds more than tall and narrow. Wide and tall holds more than wide and short. Concave or inward curves on the bottom or sides of a bottle reduce capacity. Convex or outward curves on the sides of a bottle increase capacity.]

6. What might manufacturers think about when designing the shape of a bottle? [how easy it is to grasp, how well it pours, fooling the eye into thinking there is more of the product than there really is, etc.]
7. What shape does the *Super Bottle* have? Why do you think this shape gives it the greatest capacity?
8. How does the *Super Bottle* compare to the tallest bottle? ...the widest bottle?
9. Is it always true that the tallest bottle holds the most water? Why or why not?
10. Is it always true that the widest bottle holds the most water? Why or why not?
11. What are you wondering now?

Extensions

1. Fill the bottles with a different liquid or sand and measure capacity. Are the results the same? Does the capacity change, depending on the type of filling material?
2. Try the activity with different types of containers such as boxes, cans, sour cream containers, etc.

Curriculum Correlation

Language Arts

In your science journal, write about how shape—tall, short, wide, narrow, curved—affects capacity.

* Reprinted with permission from *Principles and Standards for School Mathematics*, 2000 by the National Council of Teachers of Mathematics. All rights reserved.

All Bottled Up

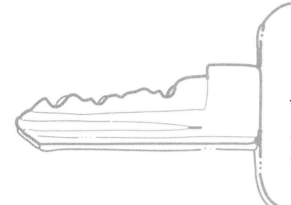

Key Question

Which bottle do you think holds the most water?

Learning Goal

Students will:

compare the capacities of four different bottles, predicting and observing the various volumes of water each can hold.

All Bottled Up

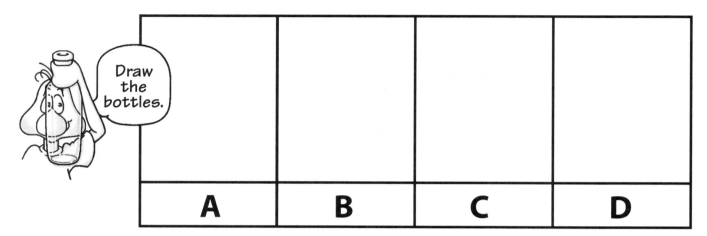

Draw the bottles.

A	**B**	**C**	**D**

Prediction

Predict the capacity of each bottle by listing them in order from the one that will hold the most to the one that will hold the least.

Holds the most ⟶ Holds the least

#1	#2	#3	#4

Actual Data

Record the amount of water that each bottle can hold.

Bottle				
mL of water				

Now list the bottles from the one that holds the most to the one that holds the least.

Holds the most ⟶ Holds the least

#1	#2	#3	#4

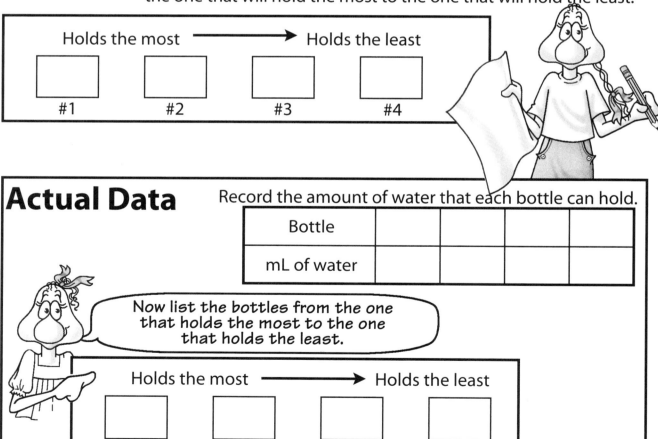

54

All Bottled Up

Enter the data from the previous page into the graph. Use the graph to find the bottle that holds the most water and the bottle that holds the least water.

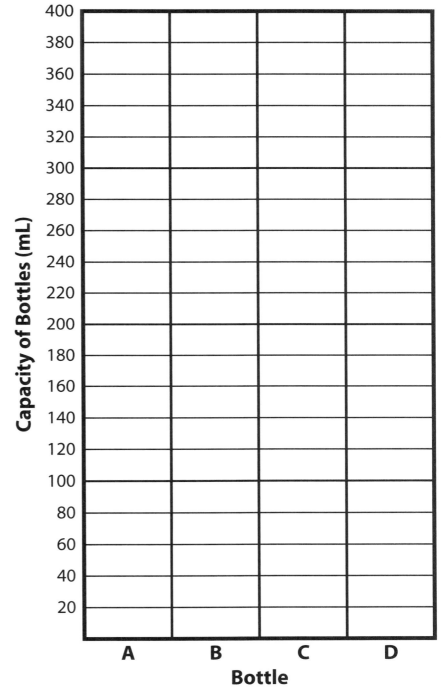

Prediction

Predict the capacity of each bottle by listing them in order from the one that will hold the most to the one that will hold the least.

Most ——————————————→ Least

#1	#2	#3	#4	#5	#6	#7	#8

Data

Record the capacity of each bottle in mL.

Bottle	a	b	c	d	e	f	g	h
mL of water								

Actual

In the frame, draw a picture of the bottle that had the greatest capacity. List the other bottles in order from greatest to least volume.

#2	#3	#4	#5	#6	#7	#8

_____
#1

It's a bird
It's a plane
It's...
SUPER BOTTLE!

All Bottled Up

Connecting Learning

1. What property of water did you observe? What other matter has this same property?

2. How did your predictions compare with the actual results?

3. Did your predictions improve as you had more experience? Why or why not?

4. What do you notice about the bottle that holds the most water? …the least water?

5. How does shape affect capacity?

Connecting Learning

6. What might manufacturers think about when designing the shape of a bottle?

7. What shape does the *Super Bottle* have? Why do you think this shape gives it the greatest capacity?

8. How does the *Super Bottle* compare to the tallest bottle? ...the widest bottle?

9. Is it always true that the tallest bottle holds the most water? Why or why not?

10. Is it always true that the widest bottle holds the most water? Why or why not?

11. What are you wondering now?

How Can We Agree ?

Topic
Volume mesurement

Key Question
How many milliliters of water does the can hold?

Learning Goals
Students will:
- estimate and measure the amount of water a can holds,
- become aware of variables that make measurement approximate, and
- reach group consensus about the can's capacity.

Guiding Documents
Project 2061 Benchmarks
- *Measurements are always likely to give slightly different numbers, even if what is being measured stays the same.*
- *Results of scientific investigations are seldom exactly the same, but if the differences are large, it is important to try to figure out why. One reason for following directions carefully and for keeping records of one's work is to provide information on what might have caused the differences.*

NRC Standard
- *Objects have many observable properties, including size, weight, shape, color, temperature, and the ability to react with other substances. Those properties can be measured using tools, such as rulers, balances, and thermometers.*

NCTM Standards 2000*
- *Understand that measurements are approximations and understand how differences in units affect precision*
- *Select and apply appropriate standard units and tools to measure length, area, volume, weight, time, temperature, and the size of angles*
- *Select and use benchmarks to estimate measurements*

Math
Estimation
Measurement
    volume/capacity (mL)
Data analysis
    range

Integrated Processes
Observing
Collecting and recording data
Comparing and contrasting
Identifying and controlling variables

Materials
For each group:
    tin can (see *Management 1*)
    graduated cylinders (see *Management 2*)
    container with water (see *Management 3*)
    paper towels

Background Information
    Measurements are approximations because any measuring unit can be further divided into smaller, more precise units. A can might hold about ½ liter, but it would be more precise to measure its capacity as 452 milliliters (thousandths of a liter) or 451,837 microliters (millionths of a liter). Scientists choose the measuring unit that best suits the purpose of the investigation.

    Human judgment also affects measurements. When is a can full? Is it when water is near the brim, at the brim, or above the brim but not overflowing? Even if there is consensus, judgment as to when the benchmark is reached will vary. How carefully is the measuring scale read? Is the person at eye level, perpendicular to the scale, so the perceived water level will not be distorted as it is at other angles? Did the person take the time to read the scale carefully?

    Technology is another variable that contributes to the approximation of measurement. How precise is the measuring tool being used? A graduated cylinder with scale increments of 100 mL is less precise than one with increments of 10 mL, or 1 mL, or 1/10 of a mL. Elementary school measuring tools are not likely to be as precisely constructed as the costly lab equipment used for scientific research.

    Students may get quite a range of results when they measure the capacity of identical cans, just as scientists may get varying results from the same experiment. As students compare their measurements, discuss how to control the variables as much as possible and perform repeated trials. Tell them that when they do this, they are going through the same processes scientists use when they seek to validate each other's work. Measurement is not so simple after all; precise measurements take great attention to detail.

Management

1. Collect identical soup or vegetable cans, one for each group of three or four students. Remove the labels so no printed measurements are visible. Make sure there are no sharp edges where the lids have been removed. If there are, cover them with masking tape.
2. Use graduated cylinders marked in 1-mL increments, if available.
3. For pouring the water, each group will need a container of water larger than the tin can. Consider using collected drink bottles.
4. Students need to work on a flat surface.

Procedure

1. Have the class brainstorm kinds of measurements (length, mass, time, etc.) Inform students that today they will be measuring capacity, the amount that a container can hold. Display a can and ask the *Key Question,* "How many milliliters of water does the can hold?"
2. Organize students into groups of three or four, distribute the activity page, and direct groups to collect a can, graduated cylinder, water container, and paper towels.
3. Instruct students to estimate and record the capacity of their group's can. Have them measure, agree on the capacity, and record.
4. Ask each group to report its measurement for the class to record in the first column of the *Class Data* table.
5. After discussing the importance of measuring things more than once, have the groups estimate and measure the volume again, report their measurements, and record in the second column.
6. Encourage students to share their observations about the data, think about what could cause differences in measurements, and decide on ways to control variables for more precise results. Emphasize that, even so, measurements are approximations. (This may be a good time to introduce the concept of tolerances.)
7. Pair the smaller groups, preferably ones with different measurements, to make four larger groups. Have the four large groups repeat the process that the smaller groups just completed and record the two trials in the *Combined Group Data* table.
8. Discuss how to agree on the capacity. Have students reflect on what they have learned about measurement.

Connecting Learning

1. How did your group's measurements compare to other groups'? How did the results make you feel? (possibly surprised at the different numbers or uncertain about which measurement was "best")
2. After studying the class data, what questions do you have? (If students take the initiative by wondering aloud and discussing at length, the rest of the questions listed may be unnecessary.)
3. Using class data, what was the range of measurements for each trial? (lowest to highest) How did the ranges of the two trials compare?
4. What could cause differences in measurements? [different ideas about the level at which the can is considered full, distortion from reading the scale at different angles, the amount of time and attention paid to reading the scale, imprecise scale increments (i.e., using a tool with 10-mL increments to measure to the nearest mL)]
5. How did your combined group come to agree on a measurement? How can the class agree on the capacity of the can? [take the median or mean average of the measurements, take extra measurements and use the most common one (mode), etc.]
6. How did the combined group data compare?
7. Is it always important to be precise in measuring? Give examples to defend your answer.
8. How could you use what you learned today for measuring length or some other kind of measurement?
9. What are you wondering now?

Extension

Repeat the same kind of activity with some other form of measurement. For example, have each group cut a piece of string 12 meters long and compare them.

Curriculum Correlation

Language Arts

Write a fictional story about a person who always measures carelessly and the problems it causes.

* Reprinted with permission from *Principles and Standards for School Mathematics,* 2000 by the National Council of Teachers of Mathematics. All rights reserved.

How Can We Agree ?

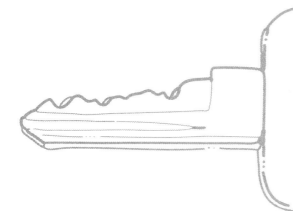

Key Question

How many milliliters of water does the can hold?

Learning Goals

Students will:

- estimate and measure the amount of water a can holds,
- become aware of variables that make measurement approximate, and
- reach group consensus about the can's capacity.

Who is correct?

How Can We Agree ?

How many milliliters of water does our can hold?

How much does it hold?

Our Group

Step One

Trial 1: Estimated capacity [ ] mL   Actual capacity [ ] mL

Trial 2: Estimated capacity [ ] mL   Actual capacity [ ] mL

Class Data

Step Two

Group	Capacity (mL)	
	Trial 1	Trial 2
1		
2		
3		
4		
5		
6		
7		
8		

Do all the groups agree on the actual capacity?

If not, why not?

Who is correct?

Combined Group Data

Step Three

Group	Capacity (mL)	
	Trial 1	Trial 2
A		
B		
C		
D		

Do all the groups agree on the actual capacity?

If not, why not?

Step Four

How can we agree on the capacity?

Final = [ ] mL

How Can We Agree?

Connecting Learning

1. How did your group's measurements compare to other groups'? How did the results make you feel?

2. After studying the class data, what questions do you have?

3. Using class data, what was the range of measurements for each trial? (lowest to highest) How did the ranges of the two trials compare?

4. What could cause differences in measurements?

How Can We Agree ?

Connecting Learning

5. How did your combined group come to agree on a measurement? How can the class agree on the capacity of the can?

6. How did the combined group data compare?

7. Is it always important to be precise in measuring? Give examples to defend your answer.

8. How could you use what you learned today for measuring length or some other kind of measurement?

9. What are you wondering now?

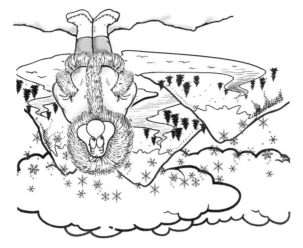

The precipitation that falls collects (accumulates) in different places. Snow collects in the mountains. In the spring or summer, it melts, and the runoff goes into lakes and rivers. Rain falls into rivers, oceans, and lakes or is absorbed into the ground. These accumulations of water are now ready to evaporate into the air and continue the cycle.

All the water on Earth is part of the water cycle. This cycle has four parts: evaporation, condensation, precipitation, and accumulation.

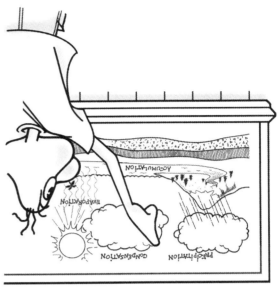

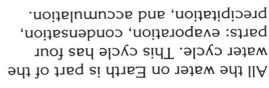

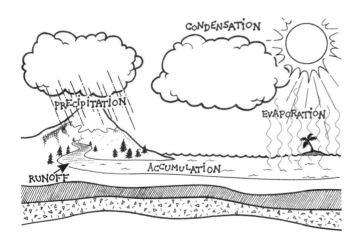

The Water Cycle

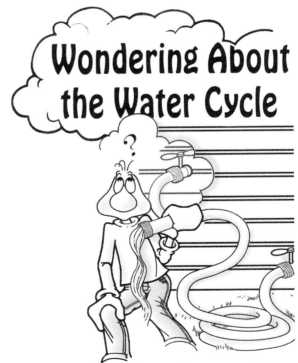

Do you ever think about the water that comes out of the tap or the hose? Where does it come from? Where has it been? Where does it go?

Evaporation occurs when the sun heats liquid water found on the Earth's surface. As the water gets hot, it changes to water vapor. This vapor, a gas, becomes part of the atmosphere.

Evaporation is one reason not to water your lawn in the hottest part of the day. Instead of going into the ground, some of the water will evaporate into the air. This won't happen as much if you water in the evening or early morning.

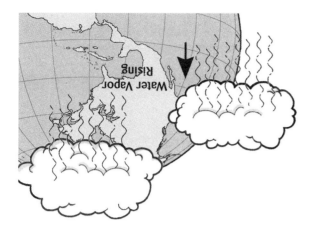

Water Vapor Rising

The water vapor that's in the air gets cooler as it rises away from the Earth. When this happens, it condenses and forms clouds.

Eventually, clouds get so much water vapor that they can't hold any more. The water then falls back to the Earth as rain, snow, sleet, hail, etc. This is precipitation.

The Mini Water Cycle

Topic
Water cycle

Key Question
What water cycle processes can we observe in a sealed plastic bag?

Learning Goal
Students will observe the processes of a miniature water cycle inside a plastic bag.

Guiding Documents
Project 2061 Benchmarks
- *When liquid water disappears, it turns into a gas (vapor) in the air and can reappear as a liquid when cooled, or as a solid if cooled below the freezing point of water. Clouds and fog are made of tiny droplets of water.*
- *Water evaporates from the surface of the earth, rises and cools, condenses into rain or snow, and falls again to the surface. The water falling on land collects in rivers and lakes, soil, and porous layers of rock, and much of it flows back into the ocean.*

NRC Standards
- *Water, which covers the majority of the earth's surface, circulates through the crust, oceans, and atmosphere in what is known as the "water cycle." Water evaporates from the earth's surface, rises and cools as it moves to higher elevations, condenses as rain or snow, and falls to the surface where it collects in lakes, oceans, soil, and in rocks underground.*
- *Materials can exist in different states—solid, liquid, and gas. Some common materials, such as water, can be changed from one state to another by heating or cooling.*

Science
Earth science
    weather
        water cycle

Integrated Processes
Observing
Recording
Relating
Generalizing

Materials
For each student:
    zipper-type plastic bag, quart size
    clear plastic cup, 3.5 oz
    permanent marker
    masking tape

Background Information
The water cycle can briefly be described as the continuous movement of the Earth's water from its surface, to the air, and back to its surface. The heat from the sun *evaporates* the water on the Earth's surface. Cool air *condenses* the water vapor into water droplets or ice crystals. They fall back to the Earth as *precipitation.*

Around 75% of the precipitation falls directly into the oceans. The rest evaporates immediately or may soak into the Earth and become part of the ground water supply. Much of the surface and ground water eventually returns to the oceans (*accumulation*), beginning the entire process over again.

A sealed plastic bag containing water models the water cycle. Warmth causes the motion of water molecules to quicken; some escape their bonds, *evaporating* into the air and becoming invisible vapor. As the water vapor cools, the molecules slow down and *condense* again into a visible liquid. Water droplets may form at the top and sides and slowly drip or slide (*precipitate*), *accumulating* at the bottom of the bag.

Management
1. Be sure to tape the bag on an angle, like a diamond, so that the sides will slant down from the top allowing the droplets to slide down and collect in the bottom corner.
2. The bags will need to be placed in a warm, sunny spot.
3. You may need to tape the cup to the inside of the bag so that the water will not spill out.

Procedure

1. Review with the students the natural water cycle, stressing the processes of evaporation, condensation, precipitation, and accumulation.
2. Tell students they will create a very simple water cycle in a closed bag and observe how water invisibly *evaporates* from the cup, like it does from oceans, *condenses* on the sides of the bag, like it does in the clouds, falls back to Earth as *precipitation,* and *accumulates* in the bottom of the bag like it does in oceans, lakes, rivers, and ground water.
3. Hand out a bag and cup to each student, or group of students. Have students place approximately two ounces of water in the cup and mark the water line with a permanent marker.
4. Instruct students to tape the cup to the inside of bag to prevent spilling.
5. Have students close the bag tightly and tape it in a warm place, tilted on an angle like a diamond. (See the student page for an illustration.)
6. Hand out the activity page and have students record what happens over a four day period.

Connecting Learning

1. What do you think will happen to the water in the cup over the four days?
2. What are the drops that have condensed on the side of the bag similar to in the real water cycle? [clouds]
3. What is the water that has collected at the bottom of the bag similar to in the real water cycle? [rivers, lakes, ground water, etc.]
4. How does the location where you placed your bag affect the amount of water that is collected in the bottom?
5. What would happen if you left your mini water cycle in a warm place for one month? Why?
6. What are you wondering now?

Extensions

1. Perform the same experiment again with salt water instead of plain water. This is one method of desalination, that is, separating salt from salt water by evaporation.
2. Perform the same experiment varying the light, color of bag, amount of water, size of bag, etc.
3. Add food coloring to the water to represent contaminants. Observe whether or not the food coloring can evaporate.

Curriculum Correlation

Language Arts

Write a story of how someone could survive with minimal water using the "water cycle in a bag" concept.

Critical thinking

Design a self-watering plant container based on the "mini water cycle."

Math

Measure the number of milliliters of water that evaporated from the cup over a four-day period.

The Mini Water Cycle

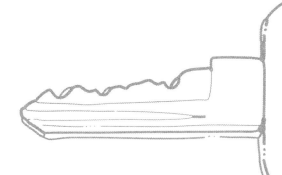

Key Question

What water cycle processes can we observe in a sealed plastic bag?

Learning Goal

Students will:

observe the processes of a miniature water cycle inside a plastic bag.

The Mini Water Cycle

Place your mini water cycle in a warm or sunny place.

1.) Draw a picture at the beginning of the experiment.

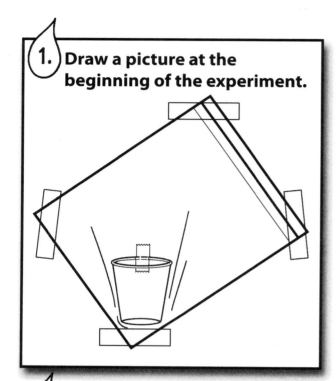

2.) Draw a picture after **two** hours.

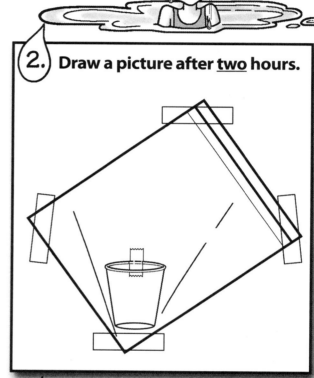

3.) Draw a picture on **Day Two**.

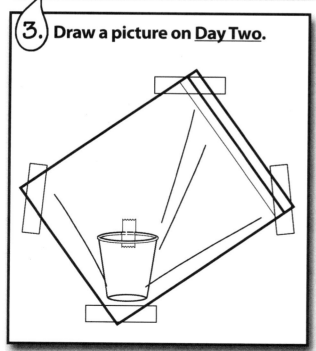

4.) Draw a picture on **Day Four**.

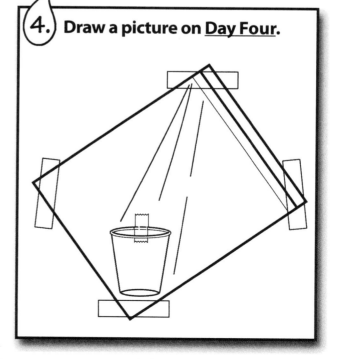

The Mini Water Cycle

Connecting Learning

1. What do you think will happen to the water in the cup over the four days?

2. What are the drops that have condensed on the side of the bag similar to in the real water cycle?

3. What is the water that has collected at the bottom of the bag similar to in the real water cycle?

4. How does the location where you placed your bag affect the amount of water that is collected in the bottom?

5. What would happen if you left your mini water cycle in a warm place for one month? Why?

6. What are you wondering now?

MOVING WATER

Topic
Water cycle

Key Question
What different forms will water take when it is heated and cooled?

Learning Goal
Students will observe water changing forms through the process of *evaporation* and *condensation*.

Guiding Documents
Project 2061 Benchmarks
- *When liquid water disappears, it turns into a gas (vapor) in the air and can reappear as a liquid when cooled, or as a solid if cooled below the freezing point of water. Clouds and fog are made of tiny droplets of water.*
- *Water evaporates from the surface of the earth, rises and cools, condenses into rain or snow, and falls again to the surface. The water falling on land collects in rivers and lakes, soil, and porous layers of rock, and much of it flows back into the ocean.*

NRC Standards
- *Materials can exist in different states—solid, liquid, and gas. Some common materials, such as water, can be changed from one state to another by heating or cooling.*
- *Water, which covers the majority of the earth's surface, circulates through the crust, oceans, and atmosphere in what is known as the "water cycle." Water evaporates from the earth's surface, rises and cools as it moves to higher elevations, condenses as rain or snow, and falls to the surface where it collects in lakes, oceans, soil, and in rocks underground.*

Science
Earth science
    weather
        water cycle

Integrated Processes
Observing
Comparing and contrasting
Generalizing

Materials
For class demonstration:
    hot plate
    teakettle
    two large metal baking pans
    one tray of ice cubes

For each student:
    student pages
    scissors
    crayons or colored pencils

Background Information
    This demonstration will bridge the gap between the concrete, real-life experience of boiling water in the kitchen to the more abstract concept of the water cycle.

    The water cycle can briefly be described as the continuous movement of the Earth's water from the oceans, to the air, and back to the ocean and land again.

    The heat from the sun *evaporates* the water from the land and the ocean. As it cools, the water vapor *condenses* into water droplets or ice crystals. They fall to the ground and the oceans in a process called *precipitation.*

    In this demonstration, the water is heated by the hot plate instead of the sun. The water vapor is invisible and may be found in the neck of the teakettle and just above the spout. The steam represents the *condensed* water vapor in the air as it cools. It is more visible as a liquid when it collects on the bottom of the metal baking pan as water droplets. As the droplets fall to the ground, comparisons can be made to precipitation.

    Around 75% of the precipitation falls back directly on the oceans. The rest evaporates immediately from the ground or different surfaces or may soak into the Earth and become part of the ground water supply. Eventually, much of the surface and ground water returns to the oceans, continuing the entire process. Thus the name *water cycle.*

Management
1. Gather a teakettle, hot plate, two large baking pans (jelly roll pans) and a tray of ice cubes for the water cycle demonstration. (You can also use an electric teapot rather than a kettle and a hot plate.)
2. Copy the student pages onto card stock. Each student needs a copy of both pages.

Procedure

1. Have students observe a steaming teakettle and describe what they see happening.
2. Ask: "What is the steam made of?" "Where did it come from?" "How did we get the liquid to go to vapor?" "How can we get it to go from a vapor to a liquid?" "Can you think of something we use to cool down a substance quickly?"
3. Fill a large pan with ice and hold it over the steaming teakettle. Observe and discuss.
4. When enough water droplets appear, shake the pan downward slightly and ask, "What form of water does this remind you of?"
5. Ask: "Can you figure out a way to collect the water drops and return them to the teakettle?"
6. Put the second baking pan on the table to catch the water dripping from the first pan.
7. Discuss how scientists have assigned names to the different changes water undergoes in the demonstration they have just seen. Discuss and define each.
   a. Evaporation: liquid water changes to a gas and goes into the atmosphere
   b. Condensation: water vapor (a gas) in the atmosphere changes back to a liquid (or solid), forming clouds
   c. Precipitation: liquid (or solid) water falls to Earth
   d. Accumulation: liquid (or solid) water collects in oceans, lakes, rivers, etc.
8. Stress that the water is undergoing a change in state (a physical change), but the water molecule always remains the same. The energy level of the water molecule affects whether it is a solid, liquid, or gas.
9. Distribute the student pages. Have students cut out the precipitation and water vapor strips on the second page. Instruct them to cut slits along the dashed lines on the first page.
10. Show them how to thread the strips through the paper to show the movement of the water.
11. Have students explain the changes the liquid water undergoes in the demonstration on the bottom of the first page.
12. Tell students to cut out the cloud and the mountain lake scene. Have them tape the cloud over the top of the pan of ice cubes and the lake scene over the cookie sheet and teakettle.
13. Discuss how the demonstration relates to the water cycle in the real world.

Connecting Learning

1. What did you notice when the water boiled? [Steam rose.]
2. As the steam went higher, what happened to it? [You could no longer see it.]
3. What happened on the underside of the tray of ice cubes? [Condensation appeared.]
4. When the drops on the tray grew large enough, what happened? [They fell downward.]
5. How does the process we modeled happen in the natural world? [It rains, water evaporates, condenses in clouds, and rains again.]
6. For each step, what state of matter is the water? [rain = liquid, evaporation = gas, condensation = liquid, precipitation = liquid]
7. Is precipitation always a liquid? [No.] Explain. [It can be a solid (snow, sleet, hail).]
8. What are you wondering now?

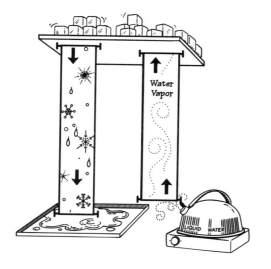

MOVING WATER

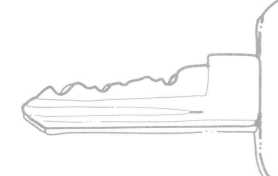

Key Question
What different forms will water take when it is heated and cooled?

Learning Goal

Students will:

observe water changing forms through the process of *evaporation* and *condensation*.

MOVING WATER

Cut out the strips on the next page. Cut slits in the picture below where there are dashed lines. Thread the strips through the slits to show how the water moves.

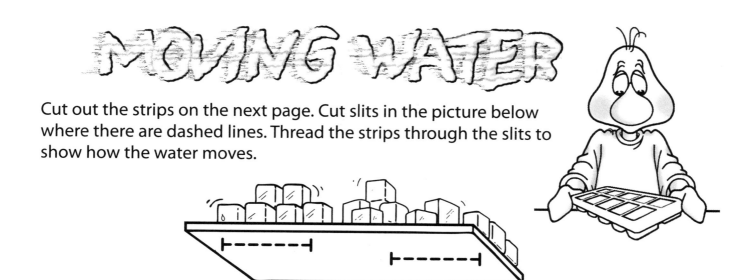

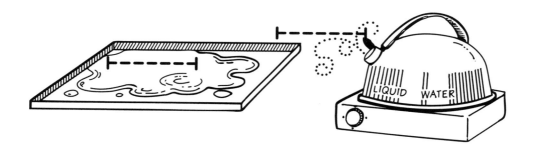

What changes does the liquid water undergo in the demonstration?

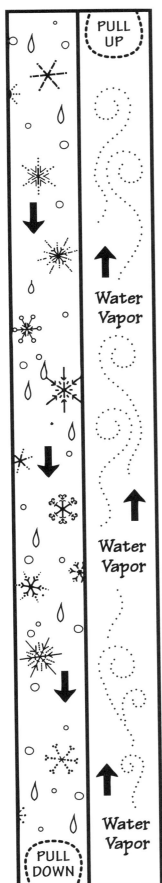

MOVING WATER

Cut out the strips and thread them through the slits on the previous page.

Cut out the cloud and the lake scene. Tape the cloud over the tray of ice cubes. Tape the lake scene over the teakettle and tray.

Connecting Learning

1. What did you notice when the water boiled?

2. As the steam went higher, what happened to it?

3. What happened on the underside of the tray of ice cubes?

4. When the drops on the tray grew large enough, what happened?

5. How does the process we modeled happen in the natural world?

6. For each step, what state of matter is the water?

7. Is precipitation always a liquid? Explain.

8. What are you wondering now?

Moving Raindrops

Topic
Water cycle

Key Question
What different forms will water go through as it moves through the water cycle?

Learning Goal
Students will construct a model that depicts the water cycle.

Guiding Documents
Project 2061 Benchmarks
- *When liquid water disappears, it turns into a gas (vapor) in the air and can reappear as a liquid when cooled, or as a solid if cooled below the freezing point of water. Clouds and fog are made of tiny droplets of water.*
- *Water evaporates from the surface of the earth, rises and cools, condenses into rain or snow, and falls again to the surface. The water falling on land collects in rivers and lakes, soil, and porous layers of rock, and much of it flows back into the ocean.*

NRC Standards
- *Materials can exist in different states—solid, liquid, and gas. Some common materials, such as water, can be changed from one state to another by heating or cooling.*
- *Water, which covers the majority of the earth's surface, circulates through the crust, oceans, and atmosphere in what is known as the "water cycle." Water evaporates from the earth's surface, rises and cools as it moves to higher elevations, condenses as rain or snow, and falls to the surface where it collects in lakes, oceans, soil, and in rocks underground.*

Science
Earth science
  weather
    water cycle
Physical science
  states of matter

Integrated Processes
Observing
Inferring
Applying

Materials
For each student:
  one paper fastener
  one jumbo craft stick
  crayons or colored pencils
  scissors
  student pages (see *Management*)

Background Information
This model of the water cycle will help to reinforce the continuous movement of water through the various processes. The water is illustrated in various forms—snow, sleet, hail, and rain—as it falls as precipitation, accumulates, evaporates, and condenses.

Management
Copy both student pages on card stock.

Procedure
1. Ask the *Key Question* and state the *Learning Goal*.
2. Discuss how the water cycle occurs on Earth with *evaporation* of water from the ocean, lakes, and the Earth's soil. The cooling of the water vapor, called *condensation*, forms clouds and fog. Review the terms *precipitation* and *accumulation* of water in lakes, rivers and streams, and finally, discuss how water moves down through the ground into the "ground water." Explain how some of the water eventually ends up in the ocean.
4. Stress that the *same* water molecule has gone through millions of changes over the years. The same water molecules that they drink today may have been in the Nile River when pharaohs ruled Egypt or part of a glacier during the last Ice Age.
5. Discuss how the water moves through the cycle, changing form from a liquid to a solid or gas (vapor), but still retaining the same ingredients of water.
6. Take students through the process of assembling their water cycle wheels.
   a. Color both wheels and cut out the wheels and the squares labeled *cut out*.
   b. Connect the two wheels by placing a paper fastener through the dot in the center of each circle. Be sure that the wheels can rotate freely.
   c. Once the wheels are connected, tape a jumbo craft stick to the back of the top wheel to act as a handle.

Connecting Learning

1. How many different changes does the liquid water in the ocean go through when moving through the water cycle? Describe these changes.
2. Is it possible for the same water molecule to be in a liquid, a gas (vapor), and a solid (ice crystal) form while going through the phases of the water cycle?
3. What causes a solid to change to a liquid and then to a vapor or gas?
4. What causes a gas or vapor to change to a liquid and then a solid?
5. What would happen if the temperature never got above 32°F or 0°C? [There would be no liquid water.] If the temperature never went below 32°F or 0°C, how would this affect life on Earth? [There would be no ice, snow, glaciers, etc.]
6. How is this model like the real water cycle? How is it different?
7. What are you wondering now?

Extensions

1. Try making a miniature water cycle by putting a closed jar with a small amount of water in it in a sunny location.
2. Place three jars in different temperature locations to note the different amounts of water vapor that collect in each jar.

Curriculum Correlation

Language Arts
Write a story from the perspective of a little rain drop or water molecule explaining where it's been or where it is going.

Literature
Locker, Thomas. *Water Dance*. Harcourt Brace & Co. Orlando. 1997.
Water speaks of its existence in such forms as storm clouds, mist, rainbows, and rivers. Includes factual information on the water cycle.

Magic School Bus Wet All Over, The: A Book About the Water Cycle. Scholastic, Inc. New York. 1996.
Ms. Frizzle's class finds out why it rains by turning into raindrops. Join them as they evaporate, condense, rain, and make their way back to the ocean... only to evaporate all over again!)

McKinney, Barbara Shaw. *A Drop Around the World*. Dawn Publications. Nevada City. 1998.
Presents the water cycle through the journey of a raindrop around the world, in the sky, on land, underground, and in the sea, in its liquid, solid, and vapor forms, as it supports life everywhere.

Internet Connections

http://www.windows.ucar.edu
Windows to the Universe
Click on Our Planet
Select Water (The Hydrosphere!)
Click on water cycle to read about the processes involved and to view another graphical model. Site contains beautiful photographs of real-world connections.

MOVING RAINDROPS

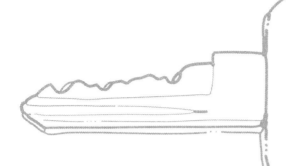

Key Question

What different forms will water go through as it moves through the water cycle?

Learning Goal

Students will:

construct a model that depicts the water cycle.

MOVING RAINDROPS

Color the wheel below and cut it out. Also cut out each of the four rectangles.

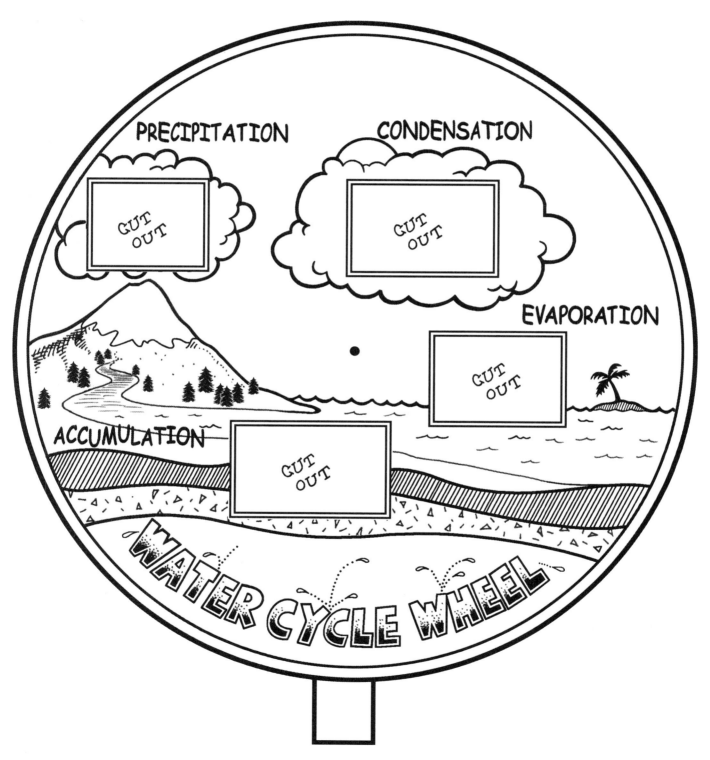

MOVING RAINDROPS

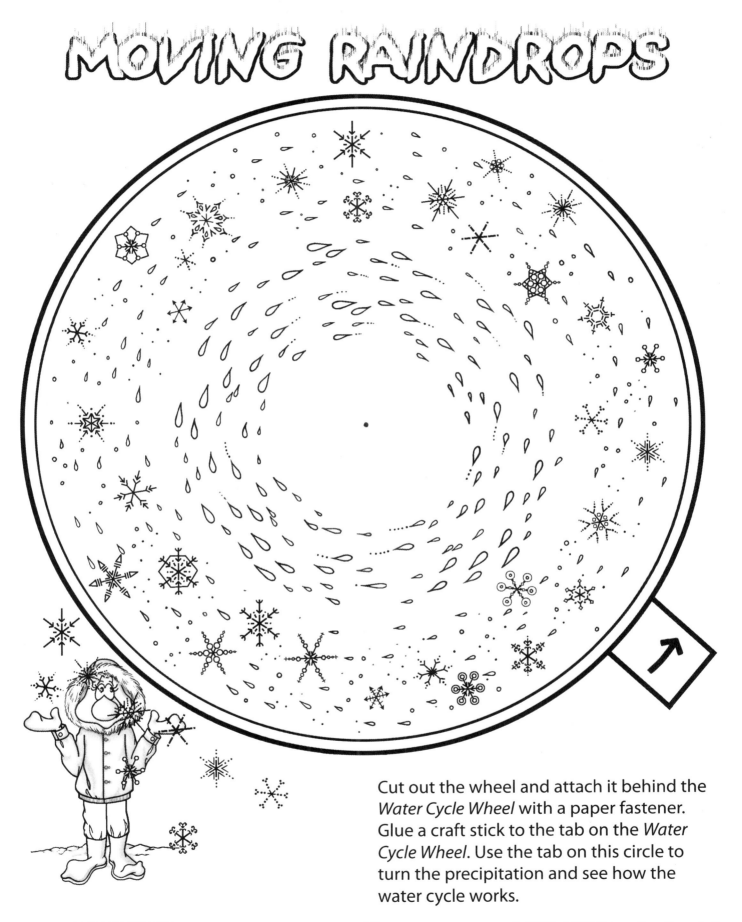

Cut out the wheel and attach it behind the *Water Cycle Wheel* with a paper fastener. Glue a craft stick to the tab on the *Water Cycle Wheel*. Use the tab on this circle to turn the precipitation and see how the water cycle works.

MOVING RAINDROPS

Connecting Learning

1. How many different changes does the liquid water in the ocean go through when moving through the water cycle? Describe these changes.

2. Is it possible for the same water molecule to be in a liquid, a gas (vapor), and a solid (ice crystal) form while going through the phases of the water cycle?

3. What causes a solid to change to a liquid and then to a vapor or gas?

4. What causes a gas or vapor to change to a liquid and then a solid?

Connecting Learning

5. What would happen if the temperature never got above 32°F or 0°C? If the temperature never went below 32°F or 0°C, how would this affect life on Earth?

6. How is this model like the real water cycle? How is it different?

7. What are you wondering now?

85

HANGING OUT TO DRY

Topic
Evaporation

Key Question
Which location will dry clothes the quickest?

Learning Goals
Students will:
- compare how well wet paper towel "shirts" dry in various locations over time,
- recognize that the liquid water is changing to water vapor—a gas, and
- identify temperature as a factor that affects how quickly water evaporates.

Guiding Documents
Project 2061 Benchmarks
- *When liquid water disappears, it turns into a gas (vapor) in the air and can reappear as a liquid when cooled, or as a solid if cooled below the freezing point of water. Clouds and fog are made of tiny droplets of water.*
- *Keep records of their investigations and observations and not change the records later.*

NRC Standards
- *Materials can exist in different states—solid, liquid, and gas. Some common materials, such as water, can be changed from one state to another by heating or cooling.*
- *Plan and conduct a simple investigation.*
- *Employ simple equipment and tools to gather data and extend the senses.*

NCTM Standards 2000*
- *Select and apply appropriate standard units and tools to measure length, area, volume, weight, time, temperature, and the size of angles*
- *Collect data using observations, surveys, and experiments*
- *Propose and justify conclusions and predictions that are based on data and design studies to further investigate the conclusions or predictions*

Math
Measurement
 elapsed time
 temperature
Data analysis
 averages

Science
Physical science
 physical change
Earth science
 weather
 water cycle
 evaporation

Integrated Processes
Observing
Predicting
Comparing and contrasting
Collecting and recording data
Inferring
Communicating

Materials
For each group:
 2 plastic drink bottles (see *Management 2*)
 string
 paper towels (see *Management 3*)
 shirt template
 scissors
 thermometer

For the class:
 pie tin or similar container
 measuring spoon or portion cup
  (see *Management 5*)
 location cards (see *Management 6*)

For each student:
 recording journal (see *Management 7*)

Background Information
 Matter on Earth commonly exists in one of three states—solid, liquid, or gas. Water is a material with which students are familiar in all three states. They have all had experience with liquid water, ice, and steam or water vapor. This activity addresses the change in the states of matter, specifically water, that is caused by an increase in temperature. The water that is put on the paper towel clothes evaporates into a gas—water vapor. Students should find that the best place to put their clotheslines is where there is a warm temperature and some air movement.

Management

1. Students will need to be divided into six groups—one for each of the locations to be tested.

2. For the clotheslines, collect two plastic drink bottles (empty or full) per group (20-oz soda or water bottles work well). Cut string into 35-cm lengths. The string can be tied to the tops of the bottles, which are then spread apart to form a clothesline. If the bottles are empty, you will need to fill them with water or sand so that they will hold up the shirt.

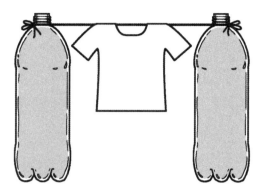

3. Use brown school-issued or blue industrial-type paper towels instead of white cloth-like ones bought in the grocery store. They are easier for students to cut, and the color change is easier for students to observe as they are drying.

4. Copy the shirt template page onto card stock and cut it in half. Each group needs one template.

5. Experiment with the paper towels you will be using beforehand so that you know about how much water each "shirt" will hold when completely saturated. Provide a measuring spoon, portion cup, etc., that will hold the correct amount of water. Have students pour this amount of water over their shirts while they are in a pie tin or similar container and swish the water around until it is completely absorbed by the paper towel. This will ensure that all groups begin with the same amount of water in their "shirts."

6. Copy the page of location cards and cut them apart. Have specific locations in mind for each card so that you can direct groups as necessary. Rewrite any locations that will not work with similar ones that will. Put the cards in a bag or cup so that each group can draw one.

7. To make the recording journals, copy the two journal pages front to back and fold them in half. If this is not possible, copy the two pages, cut them in half, stack the pages in order, and staple them along the left edge.

Procedure

1. Ask students if they can think of a time that they got their clothes wet (spilled water, ran through sprinklers, leaned up against something wet, etc.). Invite a few students to share their stories and what they did about the wet clothing.

2. Solicit students' ideas of what they would do if they had a wet spot on their shirts, but could not change clothes. Write their suggestions on the board (stand in the sun, blow on the wet spot, use a hair dryer, etc.). Look at the suggestions and see what, if anything, they have in common. Hopefully, several will involve increasing the temperature and/or air flow to the wet spot.

3. Challenge students to describe what is going on when a wet spot on a shirt dries. The water is "disappearing," but where does it go? Why do we add heat and/or air to make things dry faster? [A physical change is taking place. The liquid water is changing state to water vapor, a gas. It evaporates into the air, and this evaporation is speeded by adding heat and air circulation.]

4. Tell students that they will be exploring how the temperature and location affect how quickly clothes can dry. Explain that each group will have a different location in which they will set up a "clothesline" on which they will hang a wet paper towel "shirt."

5. Have students get into groups and distribute scissors, one paper towel, and a shirt template to each group. Have them cut out the template, fold the paper towel, place the top edge of the shirt template along the fold, trace around the shirt, and cut it out.

6. Hand around the bag with the location cards and allow each group to select one card. Give them thermometers and the materials for their clotheslines and show them how to tie the string to the bottles and spread them apart (see *Management 2*). Allow time for the groups to set up their clotheslines in the appropriate locations. Have them also set out their thermometers at this time.

7. When all groups have their clotheslines set up, have them bring their paper towel shirts to the location where you have the water and thoroughly saturate the paper towels (see *Management 5*).

8. Distribute the recording journals. Instruct students to hang the shirts over their clotheslines and record the time and temperature in their journals. (Be sure that the thermometers have been allowed to stabilize before students take readings.) Allow time for groups to complete the first recording page with detailed descriptions of their clotheslines' locations and the conditions of their shirts.

9. Distribute the student page. Have each group share the location of its clothesline. Write these on the board for reference. Instruct students to record their guesses/predictions as to which locations will produce the driest shirts by ranking the locations from most to least dry.

10. Have groups make three additional observations of their shirts. Space the observations 10 to 30 minutes apart, depending on weather conditions. (If it is a cloudy day, there is high humidity, etc., the observations should be farther apart to allow time for the shirts to dry.) At each observation, students should record the time and temperature, as well as their observations about the dryness of the shirt.

11. After the third observation, have groups bring their clotheslines, shirts, and thermometers inside and complete the summary of results in their journals.

12. Have each group write a different letter somewhere on its shirt so that they will not get mixed up. As a class, determine the order of the shirts from most to least dry.

13. Have students record the actual order in the second portion of the table on the student page. Have each group share the average temperature for its location with the class so the information can be recorded.

14. Allow time for students to answer the questions on the page, then come together for a time of class discussion and sharing.

Connecting Learning

1. Describe your group's location. Did you think your shirt would dry quickly, slowly, or somewhere in between at your location? Why?

2. Which locations had the driest shirts at the end of our observations? What do these locations have in common?

3. Which locations had the wettest shirts at the end of our observations? What do these locations have in common?

4. How did the actual results compare to your predictions?

5. Based on our data, is there a relationship between a location's temperature and how fast a shirt dries? How do you know?

6. What other factors could have influenced our data? [humidity, wind]

7. What state of matter was the water you put on the shirt? [liquid]

8. What is happening to the water in the shirt as it dries? [The water is evaporating into the surrounding air. It is changing from a liquid to a gas.]

9. Can you see the evaporation? [No.] How do you know it is happening? [The shirt dries. The water has to go somewhere.]

10. What kind of change does this represent? [physical] How do you know? [The water does not change properties, just state.]

11. What are you wondering now?

Extensions

1. Design an experiment that would control all variables except temperature to see how strong a role it plays in evaporation rate.

2. Challenge students to design a method to quickly dry a wet shirt without using any electrical tools (hair dryer, fan, etc.).

* Reprinted with permission from *Principles and Standards for School Mathematics*, 2000 by the National Council of Teachers of Mathematics. All rights reserved.

HANGING OUT TO DRY

Key Question

Which location will dry clothes the quickest?

Learning Goals

Students will:

- compare how well wet paper towel "shirts" dry in various locations over time,

- recognize that the liquid water is changing to water vapor—a gas, and

- identify temperature as a factor that affects how quickly water evaporates.

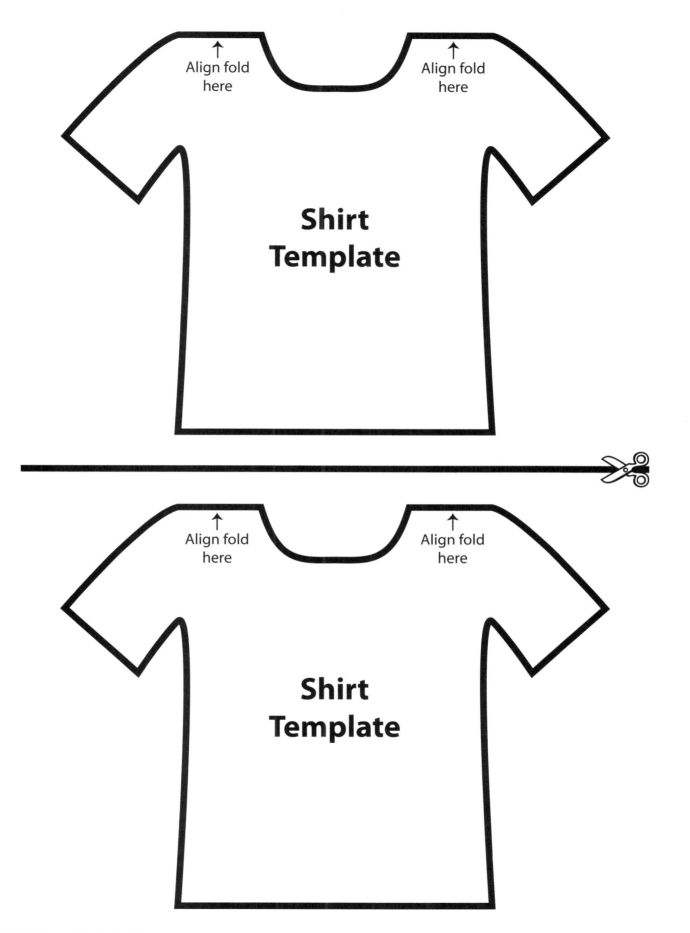

Shirt Template

Align fold here

Align fold here

Shirt Template

Align fold here

Align fold here

Location Cards

HANGING OUT TO DRY

Observation Journal

Observation Three

Time:

Temperature:

Description of shirt:

Summary of Results

Elapsed time start to finish:

Average temperature at our location:

Comparison of shirt from start to finish:

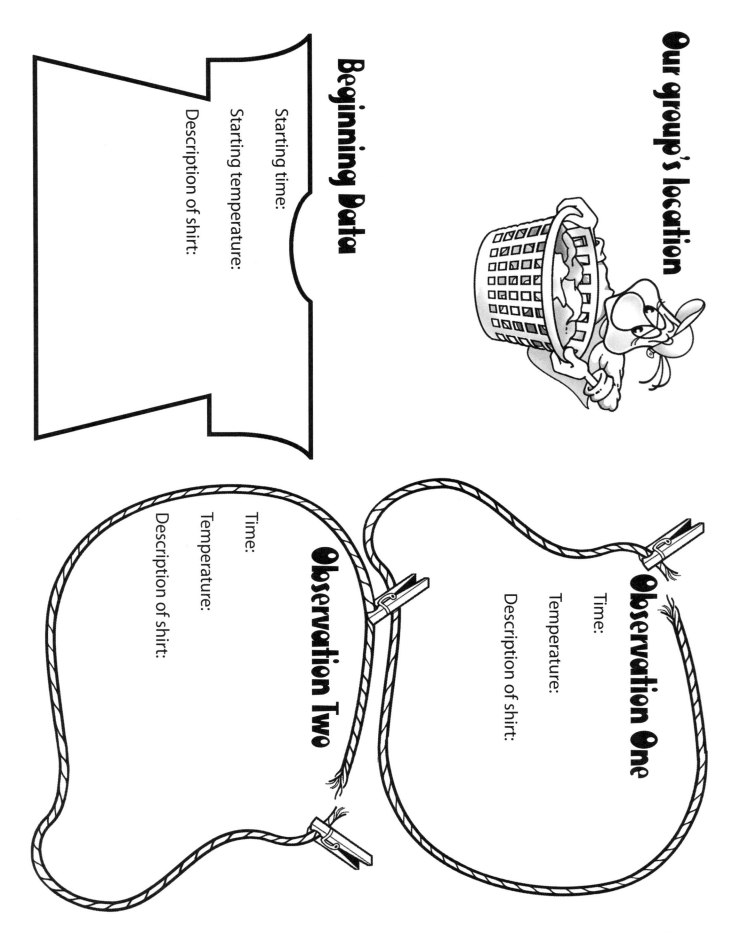

Our group's location

Beginning Data

Starting time:

Starting temperature:

Description of shirt:

Observation One

Time:

Temperature:

Description of shirt:

Observation Two

Time:

Temperature:

Description of shirt:

HANGING OUT TO DRY

Which locations do you think will have the driest shirts at the end of the observations? Write the names of the locations in the table in order from most dry to least dry.

Prediction		Actual	Average Temperature
	Most Dry		
	Least Dry		

Record the actual order of the shirts from most to least dry. Record the average temperature of each location.

1. How do your guesses/predictions compare to the actual results?

2. Does there appear to be a relationship between the temperature of a location and how dry the shirt was? Explain.

3. What other factors could have had an impact on how quickly the shirts dried?

4. What is happening to the water that was in the shirts? How do you know?

HANGING OUT TO DRY

Connecting Learning

1. Describe your group's location. Did you think your shirt would dry quickly, slowly, or somewhere in between at your location? Why?

2. Which locations had the driest shirts at the end of our observations? What do these locations have in common?

3. Which locations had the wettest shirts at the end of our observations? What do these locations have in common?

4. How did the actual results compare to your predictions?

5. Based on our data, is there a relationship between a location's temperature and how fast a shirt dries? How do you know?

Connecting Learning

6. What other factors could have influenced our data?

7. What state of matter was the water you put on the shirt?

8. What is happening to the water in the shirt as it dries?

9. Can you see the evaporation? How do you know it is happening?

10. What kind of change does this represent?

11. What are you wondering now?

MOVING MOLECULES

Topic
Evaporation

Key Question
How does the amount of surface area affect how quickly water will evaporate?

Learning Goal
Students will determine how surface area affects evaporation by measuring the water in different rectangular containers over five days.

Guiding Documents
Project 2061 Benchmarks
- *When liquid water disappears, it turns into a gas (vapor) in the air and can reappear as a liquid when cooled, or as a solid if cooled below the freezing point of water. Clouds and fog are made of tiny droplets of water.*
- *Organize information in simple tables and graphs and identify relationships they reveal.*
- *Keep records of their investigations and observations and not change the records later.*
- *Things change in steady, repetitive, or irregular ways—or sometimes in more than one way at the same time. Often the best way to tell which kinds of change are happening is to make a table or graph of measurements.*

NRC Standards
- *Use appropriate tools and techniques to gather, analyze, and interpret data.*
- *Think critically and logically to make the relationships between evidence and explanations.*
- *Use mathematics in all aspects of scientific inquiry.*

NCTM Standards 2000*
- *Collect data using observations, surveys, and experiments*
- *Represent data using tables and graphs such as line plots, bar graphs, and line graphs*
- *Understand such attributes as length, area, weight, volume, and size of angle and select the appropriate type of unit for measuring each attribute*
- *Select and apply appropriate standard units and tools to measure length, area, volume, weight, time, temperature, and the size of angles*
- *Propose and justify conclusions and predictions that are based on data and design studies to further investigate the conclusions or predictions*

Math
Measurement
    area
    volume
Whole number operations
Graphing
    bar graph
    line graph

Science
Earth science
    weather
        water cycle
            evaporation

Integrated Processes
Observing
Predicting
Comparing and contrasting
Collecting and recording data
Identifying variables
Interpreting data
Drawing conclusions

Materials
6 rectangular containers of different sizes
    (see *Management 1*)
6 metric rulers
6 small sticky notes
6 graduated cylinders
Water
Crayons or colored pencils
Student pages

Background Information
    Water can exist in three states—as a solid, a liquid, or a gas. Evaporation is a physical change in which a liquid becomes a gas. This change is affected by several variables including temperature, humidity, surface area, and wind. This activity concentrates on the effect of surface area on evaporation.
    When water molecules evaporate, they break free from the surface of the water. The number of water molecules that are able to break free into the air is partly determined by the surface area of the water. The greater the surface area, the more opportunity for water molecules to break away as vapor.

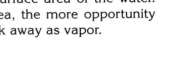

99

Management

1. Collect six different rectangular containers with perpendicular sides such as milk cartons with the tops cut off, baking pans, plastic boxes, etc. The surface areas should vary as widely as possible. The containers must be able to hold water (i.e., not cardboard).

2. If you would like a control group, collect two of each container. Cover one of each pair and compare the amount of water evaporated.

3. Divide the class into six groups. Each group will measure one container.

4. Students will determine surface area by measuring the length and width of the container's opening and multiplying. An alternative is to trace the opening on centimeter grid paper and count the squares.

5. Be aware that volume measurement can be problematic. Every time the water is measured, some drops will adhere to the container and graduated cylinder, so there will be loss not due to evaporation. Students need to avoid spilling water when pouring; place a cookie sheet underneath to catch unintended spills. To clearly observe the water level, allow bubbles to dissipate before measuring.

6. In addition to the bar graph, some classes may wish to complete the line graph. In a humid climate, water evaporates more slowly and the lines will likely be close together. The more rapid evaporation that takes place under hot, dry conditions generally yields more dramatic differences.

7. In order to have five consecutive days of measurements, start this activity on a Monday. Allow more time on the introductory and concluding days; the other days require only a few minutes to measure and record data.

First day

1. Ask, "Have you ever seen a puddle that wasn't there the next day? What happened to the water?" [It evaporated.] "Do you think the area the puddle covers affects how much water evaporates in a day?" Explain that the class will be investigating how surface area affects evaporation.

2. Give each group a container, a metric ruler, and a sticky note. Have them measure the length and width of the container's opening and calculate the area.

3. Tell groups to record the area on the sticky note, attach it to the container, and bring the container up front. Together, order the containers from smallest to largest surface area and label A (smallest) through F (largest).

4. Distribute the first activity page and have students record the surface areas of all the containers.

5. Discuss the *Key Question* and instruct students to make a prediction about how surface area will affect evaporation.

6. Give each group a graduated cylinder and have them identify the increments on the scale. Are they 1 mL, 2 mL, 5 mL, or 10 mL? Tell students to measure and pour 250 mL of water into their containers. Put the containers on a level surface not affected by temperature changes or air currents in the room.

Subsequent days

1. At the same time each day, have each group carefully pour the water in the container into the graduated cylinder, measure the volume, and return the water to the container. Ask groups to share their data for others to record.

2. On the last day, direct students to calculate and record the total milliliters evaporated.

3. Distribute the bar graph and tell students to illustrate the data. Have them look for patterns relating surface area to evaporation and record their conclusions.

4. Analyze possible errors by discussing what variables might have caused inaccurate results. [improperly reading measurement scale, water lost when poured from container to container, some containers left in the path of heated or cooled air, spills, etc.] Discuss ways to improve measurement and better control variables.

5. Urge students to think of a better way to test this variable (see *Extensions*).

6. To compare the loss of water from day to day, give capable students the line graph to complete.

Connecting Learning

1. In what ways can we make measurements more reliable? [identify the scale increments, read the scale at eye level, carefully estimate to the nearest mL, keep water from spilling, measure at the same time each day, etc.]

2. What was your prediction about the effect of surface area on evaporation? How did your prediction compare to the results?

3. What is happening when water evaporates? [It is changing state from a liquid to a gas.]

4. Does surface area affect evaporation? Use the data to support your conculsions. [The greater the surface area, the more quickly water evaporates.]

5. What variables other than surface area could affect evaporation rates? [temperature, humidity, wind, etc.]

6. What do you notice about the lines in the line graph? [Each line forms a relatively smooth slope, meaning there is a steady rate of evaporation for a given amount of surface area.]

7. How could you use the data from this activity to estimate how much water evaporates daily from an aquarium? ...a bathtub? ...a swimming pool? (Measure and calculate the surface area of the

object. Using the dimensions of the container with the largest surface area as one unit, figure how many units it would take to cover the object and multiply by the number of milliliters lost from the largest container in one day.)

8. What are you wondering now?

Extensions

1. Have students add another row to the table and calculate the *average loss per day* (total mL evaporated ÷ 5 days).
2. Use mass, rather than volume, to measure evaporation. With a scale, measure the mass of the container with 250 mL of water, then measure again each succeeding day. Since no pouring is needed, the results will likely be more accurate.
3. Use the depth of water to measure evaporation.
4. Try this activity with circular containers, determining the surface area by tracing the opening on centimeter grid paper.
5. Calculate the surface area of an aquarium, bathtub, or swimming pool and use it to estimate how much water is lost to evaporation.
6. Repeat this experiment with other liquids such as rubbing alcohol.

* Reprinted with permission from *Principles and Standards for School Mathematics*, 2000 by the National Council of Teachers of Mathematics. All rights reserved.

Key Question

How does the amount of surface area affect how quickly water will evaporate?

Learning Goal

Students will:

determine how surface area affects evaporation by measuring the water in different rectangular containers over five days.

MOVING MOLECULES

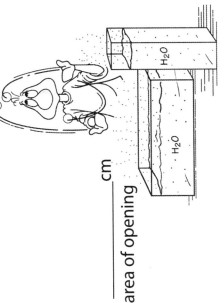

Does the area of the opening of a container affect the rate at which water will evaporate?

My Prediction:

Area of our container: $\underline{\hspace{2cm}}$ cm $\times$ $\underline{\hspace{2cm}}$ cm $=$ $\underline{\hspace{2cm}}$ cm

 length of opening $\times$ width of opening $=$ area of opening

Collect and record data. Include data from the other groups.

Container	Area of opening (l x w)	Volume of water in milliliters					Beginning Volume	Ending Volume	Total Evaporated
		Day 1	Day 2	Day 3	Day 4	Day 5			
A	cm²						250 mL		
B	cm²						250 mL		
C	cm²						250 mL		
D	cm²						250 mL		
E	cm²						250 mL		
F	cm²						250 mL		

WATER, PRECIOUS WATER

M☢VING M☢LECULES

Make a bar graph of the total water evaporated from each group's container.

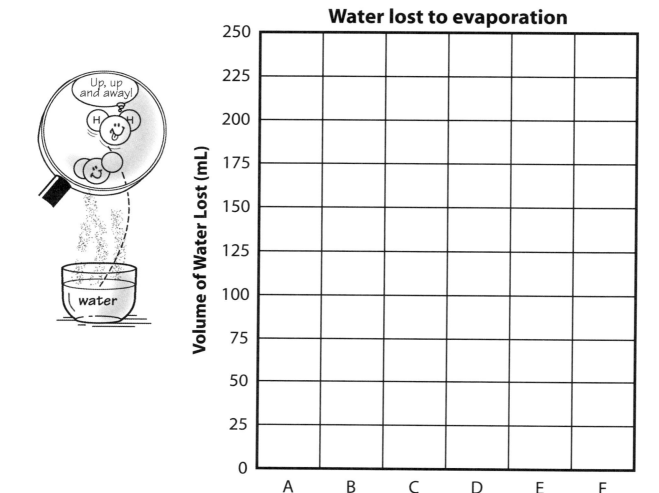

Water lost to evaporation

1. What conclusions can you draw from the data above?

2. What would be a better way to test this variable?

MOVING MOLECULES

Choose a color to represent each container and fill in the key. Construct a line graph of the data you have collected.

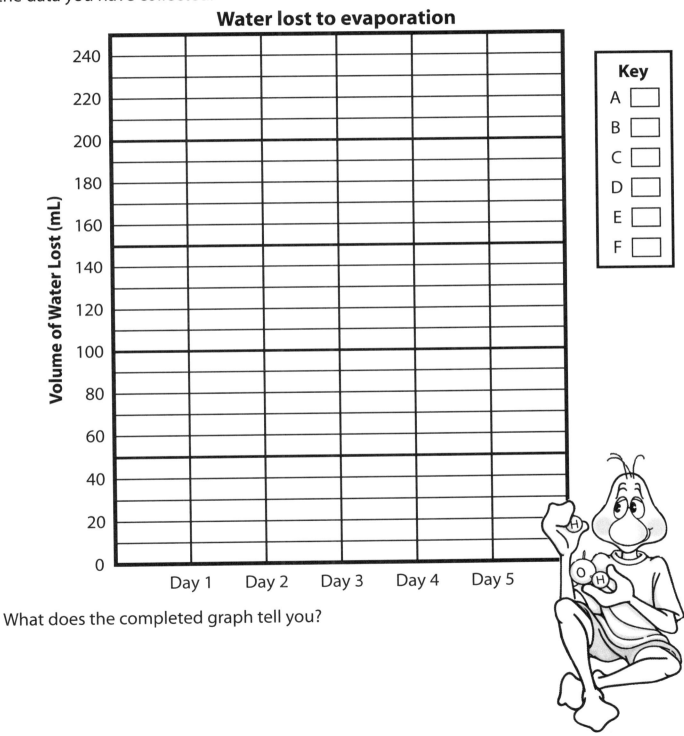

Water lost to evaporation

Key
A ☐
B ☐
C ☐
D ☐
E ☐
F ☐

Volume of Water Lost (mL)

240
220
200
180
160
140
120
100
80
60
40
20
0

Day 1   Day 2   Day 3   Day 4   Day 5

What does the completed graph tell you?

Connecting Learning

1. In what ways can we make measurements more reliable?

2. What was your prediction about the effect of surface area on evaporation? How did your prediction compare to the results?

3. What is happening when water evaporates?

4. Does surface area affect evaporation? Use the data to support your conculsions.

Connecting Learning

5. What variables other than surface area could affect evaporation rates?

6. What do you notice about the lines in the line graph?

7. How could you use the data from this activity to estimate how much water evaporates daily from an aquarium? ...a bathtub? ...a swimming pool?

8. What are you wondering now?

Concentrating on Condensation

Topic
Condensation

Key Question
What is happening when liquid appears on the outside of a cool container?

Learning Goals
Students will:
- observe the formation of condensation on the outside of a container, and
- relate this to the water cycle.

Guiding Documents
Project 2061 Benchmark
- *When liquid water disappears, it turns into a gas (vapor) in the air and can reappear as a liquid when cooled, or as a solid if cooled below the freezing point of water. Clouds and fog are made of tiny droplets of water.*

NRC Standard
- *Materials can exist in different states—solid, liquid, and gas. Some common materials, such as water, can be changed from one state to another by heating or cooling.*

NCTM Standards 2000*
- *Select and apply appropriate standard units and tools to measure length, area, volume, weight, time, temperature, and the size of angles*
- *Collect data using observations, surveys, and experiments*

Math
Measurement
    time

Science
Earth science
    weather
        water cycle
            condensation

Integrated Processes
Observing
Comparing and contrasting
Inferring
Relating

Materials
Fruit punch
Ice cubes
Cups to hold ice cubes
Shiny tin cans, labels removed
Straws or craft sticks for stirring
Soapy water and towels

Background Information
Evaporation and condensation, together with precipitation, are constantly cycling water between Earth and the atmosphere. In warmer temperatures, water molecules move more quickly, allowing them to break their bonds and escape from the surface of the liquid. Liquid water becomes water vapor, a gas, as it evaporates into the air. In cooler temperatures, water vapor molecules slow down enough to bond and condense back into water (liquid) or ice (solid). The condensation may be in the form of fog, clouds, dew, frost, or mist.

Air is invisible; so is water vapor. With evaporation, we do not see the water rise into the air. We only see the results—disappearing water puddles or drying clothes. When observing condensation, we see the reappearing water, but its source—the water vapor in the air—is invisible. This makes it more difficult for children to understand, all the more reason for them to begin observing and thinking about this process. The important idea students should gain from this activity is that water vapor is in the air. It can condense into liquid water when conditions are right, and this condensation is an important part of the water cycle.

Management
1. Students can gather tin cans, preferably with smooth surfaces, several days prior to doing the activity. Because students will be tasting the condensation that forms on the can, it is important that the cans first be washed in soapy water. They also need to be completely dry.
2. Punch is used so students can discover that the liquid that forms on the outside of the can does not come from the inside of the can.
3. Groups of two or three are recommended.

Procedure

1. Introduce the activity by asking questions such as, "Have you ever looked out the window in the morning and noticed that the grass is wet? From where do you think that water comes?" Gather student responses. Explain that condensation is what the class will be investigating today.
2. Distribute a tin can to each group. Instruct group members to breathe on the can and then rub a finger where they breathed. Ask, "What did you feel?" [dampness or wetness] "How does this compare to the tin can before breathing on it?" [It was dry.]
3. Give each student the student page and have them record what they felt.
4. Distribute the materials and allow time for students to follow the instructions and complete their observations.
5. Discuss what students learned and how condensation relates to the water cycle.

Connecting Learning

1. How did the outside of the can change? [dry, then dull film, then droplets of liquid]
2. What is this liquid? [water] From where does it come? [the water vapor in the air] How do you know? [The liquid does not taste like punch, so it did not come from the inside of the can.]
3. How did the water vapor get into the air? [Water evaporated into the air.]
4. Can you see water vapor? Explain. [No, it is a colorless gas.]
5. How do you know it is there? [You know that water vapor that condenses has to come from somewhere. You can also feel it in the air on humid days.]
6. How long did it take before the first condensation appeared on the can?
7. How long was it before the condensation turned into drops of water?
8. What is an example of condensation in the water cycle? [clouds, dew, fog]
9. What other examples of condensation can you think of? [steam from a cooking pot, breath in cold weather, etc.]
10. What are you wondering now?

Extension

Discuss when condensation becomes a problem. [mildew and mold on walls and in crevices, steam on the bathroom mirror, etc.]

Curriculum Correlation

Literature

Read the Paul Bunyan story, *The Winter of the Blue Snow*, found in a variety of books. This tall tale tells of a very cold winter where spoken words froze in the air. While water vapor usually condenses into liquid water, very cold temperatures will cause the water vapor to condense directly into ice or frost.

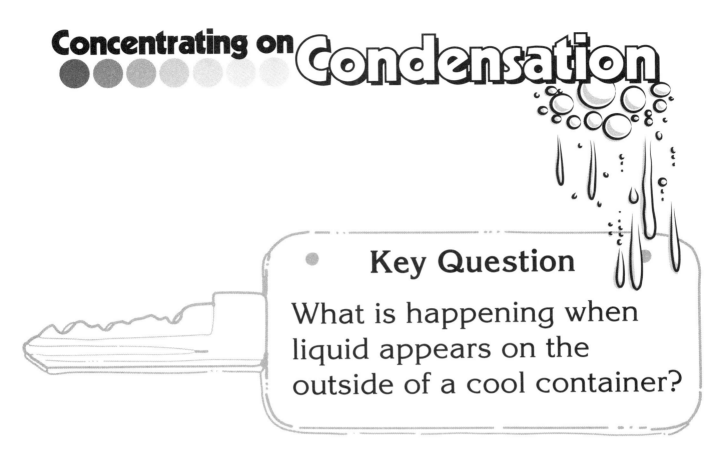

Concentrating on Condensation

Key Question

What is happening when liquid appears on the outside of a cool container?

Learning Goals

Students will:

- observe the formation of condensation on the outside of a container, and

- relate this to the water cycle.

111

Concentrating on Condensation

1. Breathe on a tin can. Rub your finger over the can. What do you feel?

2. Wash the tin can in soapy water, rinse, and dry it.
3. Fill the can two-thirds full of punch. Record the starting time.
4. Add one ice cube at a time, stirring constantly, until the surface of the can starts to dull. Record the time.
5. Continue to observe the can. Record the time when drops of liquid begin to form.

Starting Time	First Condensation Observed	Drops of Liquid Observed
_____	_____	_____

6. Collect some liquid from the outside of the can on your finger. Taste it. What does it taste like? What does this tell you?

7. What has happened in this experiment?

8. How long did it take for condensation to occur?

9. What is an example of condensation in the water cycle?

Connecting Learning

1. How did the outside of the can change?

2. What is this liquid? From where does it come? How do you know?

3. How did the water vapor get into the air?

4. Can you see water vapor? Explain.

5. How do you know it is there?

Connecting Learning

6. How long did it take before the first condensation appeared on the can?

7. How long was it before the condensation turned into drops of water?

8. What is an example of condensation in the water cycle?

9. What other examples of condensation can you think of?

10. What are you wondering now?

Water Cycle Song

Words by Suzy Gazlay

Tune: Clementine

E-vapor-a-tion, con-den-sa-tion, pre-cipit-

a-tion on my mind, a-ccumu-la-tion, the water

cy-cle, and it hap-pens all the time.

SOIL SOAKERS

Topic
Soil porosity

Key Question
How does the type of soil and its location affect the rate and depth of water penetration?

Learning Goal
Students will discover the rate at which water will soak into various soils.

Guiding Documents
Project 2061 Benchmark
- *Organize information in simple tables and graphs and identify relationships they reveal.*

NRC Standards
- *Soils have properties of color and texture, capacity to retain water, and ability to support the growth of many kinds of plants, including those in our food supply.*
- *Plan and conduct a simple investigation.*
- *Employ simple equipment and tools to gather data and extend the senses.*
- *Use mathematics in all aspects of scientific inquiry.*

NCTM Standards 2000*
- *Collect data using observations, surveys, and experiments*
- *Select and apply appropriate standard units and tools to measure length, area, volume, weight, time, temperature, and the size of angles*

Math
Measurement
    volume
    time
Graphing
Data analysis
    average

Science
Earth science
    soil porosity
    water penetration

Integrated Processes
Observing
Predicting
Collecting and recording data
Interpreting data
Generalizing

Materials
For each group:
    metric ruler
    coffee can (see *Management 3*)
    container for water (see *Management 4*)
    watch with second hand
    plastic cup, 9 oz
    graduated strip
    digging tools—metal spoons, screwdrivers, etc.

Background Information
The rate at which water will soak into soil is determined by the type of soil and its compaction. Soil that is very porous will readily accept water. Soil that has been compacted will not absorb water easily, and much of the water will either run off or evaporate. Hard-packed soils composed of clay allow very little water percolation. Soils that are already saturated do not easily absorb more water, and the excess water is lost to run off or evaporation.

Management
1. This lesson can be done over a two day period. Students can gather data on day one and do the graph and conclusion on day two.
2. Students should work together in groups of four to five.
3. Each group will need a coffee can with the top and bottom cut off. The instructions are written for large (3 lb) cans. If smaller coffee cans are used, decrease the amount of water added to 100 mL.
4. Each group will need a supply of water and a graduated cylinder or measuring cup. Two-liter soda bottles or buckets work well as portable water supplies.
5. It is best to do this activity when the soil is fairly dry. However, it is good to test a sample that is already saturated with water.
6. Copy the graduated strips onto transparency film. Tape one strip on the outside of each plastic cup to be used as a measuring device.

Procedure

1. Discuss ground water and how the water soaks (percolates) into the soil.
2. Discuss how the type of soil and its location could affect the rate at which it soaks up water.
3. Divide the class into groups and distribute the materials.
4. Hand out the first two student pages and go over the procedure. (See the first page for a detailed description.)
5. Have students make their predictions before going outside to conduct the activity.
6. Be sure that groups select soil samples that vary in compaction (e.g., sand, rocky soil, packed clay, soil recently dug up, soil frequently walked on, etc.).
7. Instruct groups to test at least three different soil locations (e.g., sand, hard-packed soil, wet soil, loose soil, etc.). Students should use a digging tool to dig up the soil for the loose-soil sample. If time allows, have each group collect data from three different soil types.
8. Monitor the groups to be sure that they are following the procedure correctly.
9. When all groups have collected their data, return to the classroom, average the data, analyze the data, and graph it. Students will have to determine the time intervals to be used on the graph based on the types of soil tested.

Connecting Learning

1. Why did the soil locations absorb water at different rates?
2. Which kind of soil absorbed water the most quickly? Why?
3. Which kind of soil absorbed water the slowest? Why?
4. How can you change soil that doesn't soak up water well into soil that absorbs water more readily?
5. How does soil get compacted?
6. Why do gardeners often dig around their plants?
7. What did you learn from doing this activity?
8. What are you wondering now?

Extensions

1. Select soil from one location, compact half of it, and compare the percolation rate of the compacted sample to the uncompacted one.
2. Plant seeds in soils of various compaction and compare germination rates and how well they grow.

* Reprinted with permission from *Principles and Standards for School Mathematics*, 2000 by the National Council of Teachers of Mathematics. All rights reserved.

SOIL SOAKERS

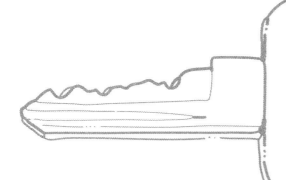

Key Question

How does the type of soil and its location affect the rate and depth of water penetration?

Learning Goal

Students will:

discover the rate at which water will soak into various soils.

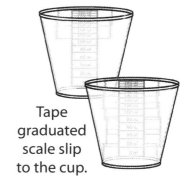

Tape graduated scale slip to the cup.

Graduated Scales
for 9-oz cups

Cover completely with tape.

240	240	240	240	240	240	240	240	240	240
220	220	220	220	220	220	220	220	220	220
200	200	200	200	200	200	200	200	200	200
180	180	180	180	180	180	180	180	180	180
160	160	160	160	160	160	160	160	160	160
140	140	140	140	140	140	140	140	140	140
120	120	120	120	120	120	120	120	120	120
100	100	100	100	100	100	100	100	100	100
80	80	80	80	80	80	80	80	80	80
60	60	60	60	60	60	60	60	60	60
40	40	40	40	40	40	40	40	40	40
20	20	20	20	20	20	20	20	20	20
mL	mL	mL	mL	mL	mL	mL	mL	mL	mL
240	240	240	240	240	240	240	240	240	240
220	220	220	220	220	220	220	220	220	220
200	200	200	200	200	200	200	200	200	200
180	180	180	180	180	180	180	180	180	180
160	160	160	160	160	160	160	160	160	160
140	140	140	140	140	140	140	140	140	140
120	120	120	120	120	120	120	120	120	120
100	100	100	100	100	100	100	100	100	100
80	80	80	80	80	80	80	80	80	80
60	60	60	60	60	60	60	60	60	60
40	40	40	40	40	40	40	40	40	40
20	20	20	20	20	20	20	20	20	20
mL	mL	mL	mL	mL	mL	mL	mL	mL	mL

SOIL SOAKERS

Procedure:

1. Choose a location for your first sample.

2. Push and rotate your coffee can into the ground about two centimeters.

3. Pour 250 mL of water into the can. If the water leaks around the edges, rotate the can until it stops leaking.

4. Time how long it takes for the soil to absorb the water. Record this data in the table.

5. Move the can to a different spot. Pick the same kind of soil. The ground should not be already wet. Repeat steps two through four.

6. Repeat for as many different soil types as you can.

7. Complete the table by determining the averages for each location.

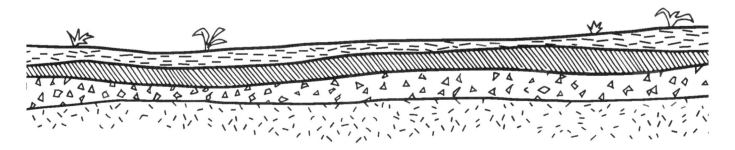

SOIL SOAKERS

How does the type of soil and its location affect the rate at which it soaks up water?

I think:

Test three different locations. Record your data.

Location	Soil Type	Time to absorb Trial One	+	Time to absorb Trial Two	=	Total of both trials	Average (Total ÷ 2)
1.			+		=		
2.			+		=		
3.			+		=		

1. Which soil location had the highest average time?

2. Which soil location had the lowest average time?

3. Rank the soil locations from the most absorbent to the least absorbent.

Most absorbent ← → **Least absorbent**

SOIL SOAKERS

Make a bar graph of the time it took each soil to absorb the water. Pick numbers for the time intervals based on your data. Label the graph appropriately.

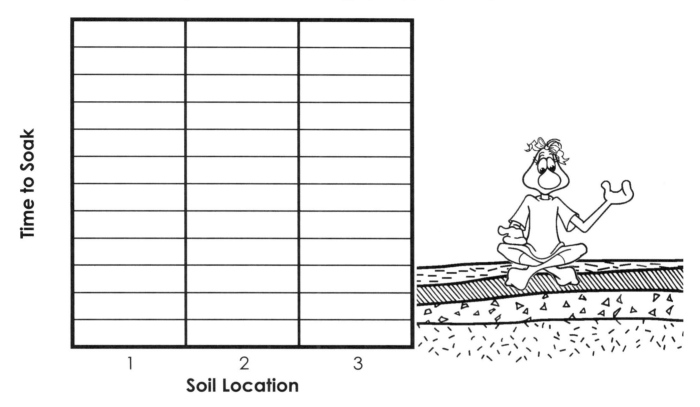

Time to Soak (vertical axis)

Soil Location (horizontal axis): 1 2 3

1. Why did the soil locations absorb at different rates?

2. Why do gardeners often dig around their plants?

3. How does soil get compacted?

4. What did you learn from doing this activity?

Connecting Learning

1. Why did the soil locations absorb water at different rates?

2. Which kind of soil absorbed water the most quickly? Why?

3. Which kind of soil absorbed water the slowest? Why?

4. How can you change soil that doesn't soak up water well into soil that absorbs water more readily?

5. How does soil get compacted?

6. Why do gardeners often dig around their plants?

7. What did you learn from doing this experiment?

8. What are you wondering now?

Topic
Erosion

Key Question
What happens to soil on a bare hillside when it rains?

Learning Goal
Students will observe the effects of erosion caused by rain on a bare hillside.

Guiding Documents
Project 2061 Benchmarks
- *Waves, wind, water, and ice shape and reshape the earth's land surface by eroding rock and soil in some areas and depositing them in other areas, sometimes in seasonal layers.*
- *Although weathered rock is the basic component of soil, the composition and texture of soil and its fertility and resistance to erosion are greatly influenced by plant roots and debris, bacteria, fungi, worms, insects, rodents, and other organisms.*

NRC Standard
- *The surface of the earth changes. Some changes are due to slow processes, such as erosion and weathering, and some changes are due to rapid processes, such as landslides, volcanic eruptions, and earthquakes.*

Math
Measurement

Science
Earth science
  erosion

Integrated Processes
Observing
Comparing and contrasting

Materials
For each group:
  cookie sheet or large shallow pan
  book or block of wood (see *Management 2*)
  300 mL mixed dirt and sand (see *Management 3*)
  200 mL water
  Styrofoam cup or paper cup, 8 oz or larger
  tape
  centimeter ruler
  crayons: green, brown, and blue
  plastic cup, 9 oz
  graduated strip

Background Information
Erosion is the carrying away of land surface by various forces. The most significant causes of erosion are wind, water, and ice. Erosion can be controlled if slopes are covered with vegetation or terraced.

Management
1. Students should work together in groups of four to six. Each group will need its own hill to work with.
2. Groups should use books or blocks of wood to prop up one end of the pans. They should be about 1 1/2" high to get the appropriate slant for runoff. If you use books, be sure to cover them with plastic wrap so that they will not be damaged by the water.
3. Be sure the sand and dirt mixture is moist enough to form into a hill.
4. Each group will need a 9-oz cup with an attached graduated strip to use as a volume measuring cup.
5. Groups will need either a Styrofoam or a paper cup that has at least an 8-oz capacity in order to hold 200 mL of water. It must be Styrofoam or paper so that they can poke holes in the bottom using a sharpened pencil.

Procedure
1. Divide the class into groups and distribute the materials.
2. Have groups form hills out of sand and dirt at one end of their pans. Have students elevate the pans using books or blocks of wood so that the hills are at the higher end.
3. Instruct students to draw pictures of their hills in the picture marked "Before" on the student page.
4. Have each group poke five holes in the bottom of the Styrofoam/paper cup using the point of a sharpened pencil and cover the holes with tape.

5. Using the calibrated cup, have groups measure out 200 mL of water and pour it into the Styrofoam/paper cup.
6. Direct students to hold their cups about 10 cm above the hills, pull the tape off the bottoms of the cups, and allow the "rain" to fall.
7. Once no more water will fall from the cup, have students draw what the pan and hill look like in the picture labeled "After the first rain." Be sure they show where the dirt is, what the hill looks like, and where the water is located.
8. Do not allow students to rebuild the hill. Have them repeat the process a second time and draw the results in the picture labeled "After the second rain."
9. Have students feel the dirt at the bottom of the pan and the dirt at the base of the hill and compare them.
10. Instruct groups to carefully pour the water from the pan back into the graduated cup without dumping the dirt. Have them record the amount of runoff and compute the difference between the original 400 mL of rain water and the total amount of runoff. (Because of the small amount of water left in the cup after each "rain," the total amount of water will not be exactly 400 mL, but this value is close enough for an estimation of the amount of runoff.)

Connecting Learning

1. What happens to the hill when rain falls?
2. How could the dirt be kept on the hill during a rain?
3. Why would it be better to keep the dirt on the hill?
4. Where did most of the dirt settle: at the base, midway, or all the way down?
5. Why do you think it settled where it did?
6. Is erosion a constructive or destructive process? Explain.
7. What are some surface features created by erosion? How does your model show these?
8. How did the dirt at the bottom of the pan feel compared to the dirt at the base of the hill? Why?
9. What would have happened to the dirt if it had rained harder?
10. How much water were you able to collect after the rain?
11. Where do you think the missing water went?
12. What are you wondering now?

Extensions

1. Repeat the experiment with a variety of soil types and compare.
2. Repeat the experiment with more holes in the bottom of the cup so that it rains "harder."

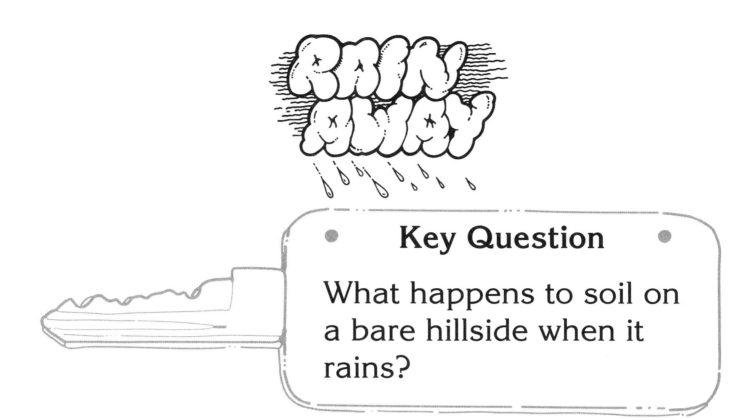

Key Question

What happens to soil on a bare hillside when it rains?

Learning Goal

Students will:

observe the effects of erosion caused by rain on a bare hillside.

Graduated Scales
for 9-oz cups

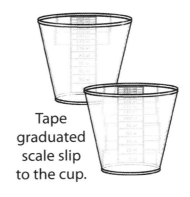

Tape graduated scale slip to the cup.

Cover completely with tape.

240	240	240	240	240	240	240	240	240	240
220	220	220	220	220	220	220	220	220	220
200	200	200	200	200	200	200	200	200	200
180	180	180	180	180	180	180	180	180	180
160	160	160	160	160	160	160	160	160	160
140	140	140	140	140	140	140	140	140	140
120	120	120	120	120	120	120	120	120	120
100	100	100	100	100	100	100	100	100	100
80	80	80	80	80	80	80	80	80	80
60	60	60	60	60	60	60	60	60	60
40	40	40	40	40	40	40	40	40	40
20	20	20	20	20	20	20	20	20	20
mL	mL	mL	mL	mL	mL	mL	mL	mL	mL

240	240	240	240	240	240	240	240	240	240
220	220	220	220	220	220	220	220	220	220
200	200	200	200	200	200	200	200	200	200
180	180	180	180	180	180	180	180	180	180
160	160	160	160	160	160	160	160	160	160
140	140	140	140	140	140	140	140	140	140
120	120	120	120	120	120	120	120	120	120
100	100	100	100	100	100	100	100	100	100
80	80	80	80	80	80	80	80	80	80
60	60	60	60	60	60	60	60	60	60
40	40	40	40	40	40	40	40	40	40
20	20	20	20	20	20	20	20	20	20
mL	mL	mL	mL	mL	mL	mL	mL	mL	mL

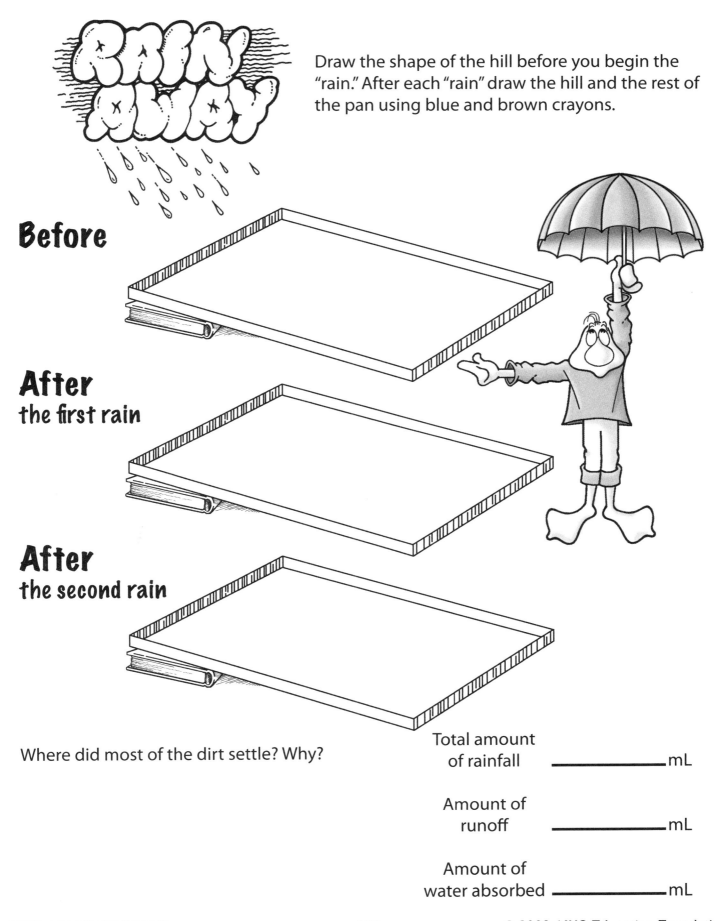

Draw the shape of the hill before you begin the "rain." After each "rain" draw the hill and the rest of the pan using blue and brown crayons.

Before

After
the first rain

After
the second rain

Where did most of the dirt settle? Why?

Total amount of rainfall	_____ mL
Amount of runoff	_____ mL
Amount of water absorbed	_____ mL

Connecting Learning

1. What happens to the hill when rain falls?

2. How could the dirt be kept on the hill during a rain?

3. Why would it be better to keep the dirt on the hill?

4. Where did most of the dirt settle: at the base, midway, or all the way down?

5. Why do you think it settled where it did?

6. Is erosion a constructive or destructive process? Explain.

Connecting Learning

7. What are some surface features created by erosion? How does your model show these?

8. How did the dirt at the bottom of the pan feel compared to the dirt at the base of the hill? Why?

9. What would have happened to the dirt if it had rained harder?

10. How much water were you able to collect after the rain?

11. Where do you think the missing water went?

12. What are you wondering now?

Topic
Erosion

Key Question
How can water erosion on a hillside be controlled?

Learning Goal
Students will observe how vegetation helps control the erosion caused by rain.

Guiding Documents
Project 2061 Benchmarks
- *Waves, wind, water, and ice shape and reshape the earth's land surface by eroding rock and soil in some areas and depositing them in other areas, sometimes in seasonal layers.*
- *Although weathered rock is the basic component of soil, the composition and texture of soil and its fertility and resistance to erosion are greatly influenced by plant roots and debris, bacteria, fungi, worms, insects, rodents, and other organisms.*
- *Make sketches to aid in explaining procedures or ideas.*

NRC Standard
- *The surface of the earth changes. Some changes are due to slow processes, such as erosion and weathering, and some changes are due to rapid process, such as landslides, volcanic eruptions, and earthquakes.*

NCTM Standards 2000*
- *Understand such attributes as length, area, weight, volume, and size of angle and select the appropriate type of unit for measuring each attribute*
- *Select and apply appropriate standard units and tools to measure length, area, volume, weight, time, temperature, and the size of angles*

Math
Measurement
    volume

Science
Earth science
    erosion

Integrated Processes
Observing
Comparing and contrasting
Collecting and recording data

Materials
For each group:
    "hillside" from *Rain Away*
    rye grass seeds
    paper cups, 3 oz
    potting soil
    Styrofoam cups from *Rain Away*
    spray bottle of water
    calibrated cup or beaker
    crayons or colored pencils

Background Information
Erosion is the wearing away of land surface by various forces. The most significant causes of erosion are wind, water and ice. Erosion can be controlled if slopes are covered with vegetation or terraced.

Management
1. This activity is intended to follow *Rain Away* and should use the same setup.
2. You will need a place where students can leave the hill setup undisturbed for a week or more until the seeds sprout.
3. Rye grass seed can be purchased in small bags from home improvement stores or nurseries. The seeds sprout quickly. If you plant the seeds on a Monday, you should be able to do the activity the following Monday.

Procedure
1. Immediately following *Rain Away*, while the soil is still damp, have students rebuild their hills.
2. Give each group a paper cup full of rye grass seeds. Instruct them to sprinkle the seeds over the entire hill as evenly as possible.
3. Have students fill the paper cups with potting soil and sprinkle a thin layer of potting soil over the seeds. Instruct them to dampen the potting soil by misting it with the spray bottle of water.
4. Have groups move their hills to a location where they will be undisturbed. Explain that they will be waiting for the grass to grow on their hills and will need to monitor their hills every day. If the soil starts to dry out over the course of the week, they are to mist it with the spray bottle of water.

5. After the hills all have a good amount of growth, about five to seven days after planting, have students carefully move their hills back to their desks. Instruct them to set up the hills as they did in *Rain Away* with the hill end elevated by a stack of books or blocks. Have them get out their materials from *Rain Away*.

6. Using the calibrated cup, have groups measure out 200 mL of water and pour it into the Styrofoam cup.

7. Direct students to hold their cups about 10 cm above the hills, pull the tape off the bottoms of the cups, and allow the "rain" to fall.

8. Once all 200 mL has fallen, have students draw what the pan and hill look like in the picture labeled "After the first rain." Be sure they show where the dirt is, what the hill looks like, and where the water is located.

9. Do not allow students to rebuild the hill. Have them repeat the process a second time and draw the results in the picture labeled "After the second rain."

10. Have students feel the dirt at the bottom of the pan and the dirt at the base of the hill and compare them.

11. Instruct groups to carefully pour the water from the pan back into the graduated cup without dumping the dirt. Have them record the amount of runoff and compute the difference between the original 400 mL of rain water and the total amount of run off.

12. Compare the results from *Rain Away* to this activity and discuss the effects the grass had on erosion.

Connecting Learning

1. What effect did covering the hill with vegetation have when it rained?
2. How did the results of *Rain Away* compare to the results you got in this activity?
3. Why is there more erosion in the hills after a fire?
4. What else can be done to lessen erosion caused by rain?
5. What are you wondering now?

Extensions

1. Terrace the "hills" and repeat the activity.
2. Plant additional vegetation on the hill, such as flowers, then repeat the activity.

* Reprinted with permission from *Principles and Standards for School Mathematics*, 2000 by the National Council of Teachers of Mathematics. All rights reserved.

Key Question

How can water erosion on a hillside be controlled?

Learning Goal

Students will:

observe how vegetation helps control the erosion caused by rain.

Draw the shape of the hill before you begin the "rain." After each "rain," draw the hill and the rest of the pan.

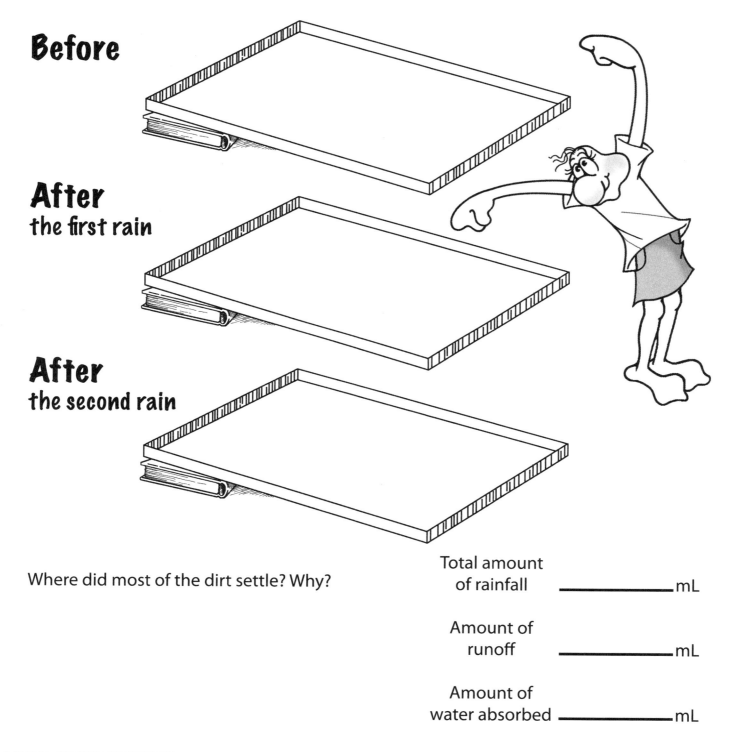

Before

After
the first rain

After
the second rain

Where did most of the dirt settle? Why?

Total amount
of rainfall _____ mL

Amount of
runoff _____ mL

Amount of
water absorbed _____ mL

Connecting Learning

1. What effect did covering the hill with vegetation have when it rained?

2. How did the results of *Rain Away* compare to the results you got in this activity?

3. Why is there more erosion in the hills after a fire?

4. What else can be done to lessen erosion caused by rain?

5. What are you wondering now?

Down the Drain

You can save about 300 gallons of water a month by watering your lawn early in the morning or in the evening.

The average person in the U.S. uses more than 100 gallons of water a day. Taking a five-minute shower instead of a bath can save as much as 20 gallons of water. That would fill 38 two-liter soda bottles.

If you turn off the water while you brush your teeth, you could save up to four gallons of water per day!

Other uses of water can add up quickly. One can of soda requires 29,000 gallons of water to produce.

Down the Drain!

Topic
Water conservation

Key Question
How much water is wasted while waiting for tap water to become warm?

Learning Goals
Students will:
- measure the amount of water wasted; and
- project potential water waste for their city, state, and country.

Guiding Documents
Project 2061 Benchmarks
- *The benefits of the earth's resources—such as fresh water, air, soil, and trees—can be reduced by using them wastefully or by deliberately or inadvertently destroying them. The atmosphere and the oceans have a limited capacity to absorb wastes and recycle materials naturally. Cleaning up polluted air, water, or soil or restoring depleted soil, forests, or fishing grounds can be very difficult and costly.*
- *Use numerical data in describing and comparing objects and events.*

NRC Standards
- *Plan and conduct a simple investigation.*
- *Use mathematics in all aspects of scientific inquiry.*
- *The supply of many resources is limited. If used, resources can be extended through recycling and decreased use.*

NTCM Standards 2000*
- *Collect data using observations, surveys, and experiments*
- *Use measures of center, focusing on the median, and understand what each does and does not indicate about the data set*
- *Develop fluency in adding, subtracting, multiplying, and dividing whole numbers*
- *Select and apply appropriate standard units and tools to measure length, area, volume, weight, time, temperature, and the size of angles*
- *Solve problems that arise in mathematics and in other contexts*

Math
Whole number operations
Data analysis
   median or mean

Measurement
   elapsed time
   volume

Science
Environmental science
   natural resources
      water conservation

Integrated Processes
Observing
Comparing and contrasting
Collecting and recording data
Analyzing data
Relating

Materials
Large bucket
Stopwatch or timer (see *Management 6*)
Liter measuring tool
Calculators, optional
Student pages
Internet access

Background Information
As the population of a community grows, the demand for water increases. Conservation of water is a vital aspect for present and future considerations. One way that water is wasted is by running the tap until the water gets warm.

Students will measure the amount of water wasted while waiting for it to get warm and then project their results onto a larger sample of water users. Many hot water tanks are located far from the point of use. Locating the tank nearer to the sinks would decrease the amount of water being wasted.

Management
1. For *Part One*, students should estimate and measure with the volume units available at home, most likely quarts or cups. They will use metric units in *Part Two*.
2. For *Part Two*, locate a sink with a hot water tap at your school.
3. Be sure the tap is cool before beginning. Wait a few minutes if it has recently been used.

4. Depending on your goals, decide whether students will record the median (middle number when the three amounts are ordered) or the mean (three amounts totaled and divided by three).
5. This activity takes two days. On the first day, allow for a short introductory period, followed by the *Part One* home assignment. The next day, plan for a longer time to complete *Part Two*.
6. A stopwatch or digital kitchen timer that counts up are best to use for timing. If these are not available, a watch or clock with a second hand may be used.

Procedure
Part One: Home
1. Ask the *Key Question,* "How much water is wasted while waiting for tap water to become warm?" Have students think about an estimate while you distribute *Part One*.
2. Read the directions together. Tell students to estimate and record how many seconds or minutes it will take for the kitchen water to become warm. Record the range of estimates in a place visible to the class.
3. Have students take the page home, locate a measuring cup or larger measuring tool to identify the measuring units they will use, and estimate the amount of water that will flow from the kitchen tap while waiting for it to become warm.
4. Instruct them to turn on the faucet and measure the time it actually took and the amount of water collected.
5. Ask them to repeat the estimates and measurements for the bathroom tap and bring the paper to class the next day.

Part Two: School
1. Begin by discussing the results of *Part One*.
2. Give students *Part Two*. Take them to the hot water tap and hold up a liter measurer, noting that it may be different than what they used at home. Have them estimate how many liters of water will be wasted in one trial.
3. Put a bucket under the faucet and turn it on. As soon as the water is slightly warm, run the tap for five more seconds and stop. Optional: have someone time how long it takes.
4. Return to the classroom and have students measure the amount of water.
5. Repeat the same procedure for two more trials. Be sure the tap has cooled between trials.
6. Identify the average to be used (median or mean). Have students determine and record the average amount of water wasted.
7. Have students research the current populations of their city, state, and country on the Internet. See *Internet Connections* for recommended websites. Alternatively, use almanacs.

8. Direct students to compute the projected water waste for their city, state, and country.
9. Discuss ways to solve the water waste problem. Contact the local water company as a resource.

Connecting Learning
Part One: Home
1. How did the times it took your kitchen and bathroom taps to produce warm water compare?
2. What were the shortest and longest times in our class?
3. How did the amount of water from your two taps compare?
4. If the two taps had different results, why do you think so? (Hint: Where is the hot water heater located in relation to the kitchen sink and the bathroom sink?)
5. Whose tap conserved the most water?

Part Two: School
1. What was the average amount of water wasted? (If you determined both averages, how did the median compare to the mean?)
2. How close were your estimates to the actual amount of water wasted?
3. What variables might affect how much water is collected? [if the tap started out cooler one time than the next time, a person's judgment about when the water has become warm, how far the tap was turned on, etc.]
4. How might you solve the problem of wasting water? [Instead of letting it go down the drain, collect it and use to water plants, etc.]
5. What are you wondering now?

Extensions
1. Investigate how much water is wasted waiting for the tap to become cool.
2. Create water conservation posters.

Internet Connections
U.S. Census Bureau
http://factfinder.census.gov/
Click on the "Population Finder" link to access population data for the country, states, counties, and cities.

H2ouse Water Saver Home
http://www.h2ouse.org
Select the "Home Tour" for statistics on water use in the home and suggestions for ways to save water in and around the home.

* Reprinted with permission from *Principles and Standards for School Mathematics*, 2000 by the National Council of Teachers of Mathematics. All rights reserved.

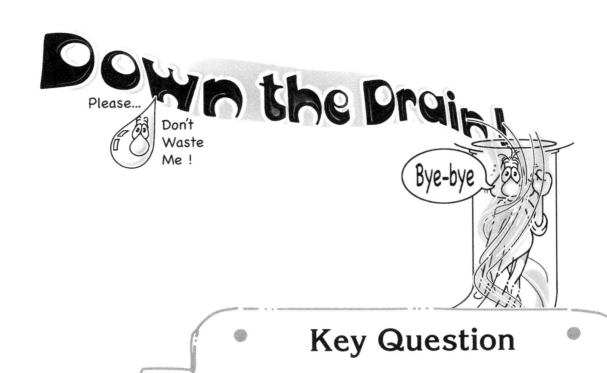

Down the Drain!

Please... Don't Waste Me !

Bye-bye

Key Question

How much water is wasted while waiting for tap water to become warm?

Learning Goals

Students will:

- measure the amount of water wasted; and
- project potential water waste for their city, state, and country.

STILL not warm...

HOT

Part One: Home

How much water is wasted while waiting for tap water to become warm?

After reading the directions below, record your estimates.
- Put the bucket under a faucet that has not been used for at least 10 minutes.
- Turn on the faucet. Time how long it takes for the water to become warm.
- Measure the amount of water collected in the bucket.

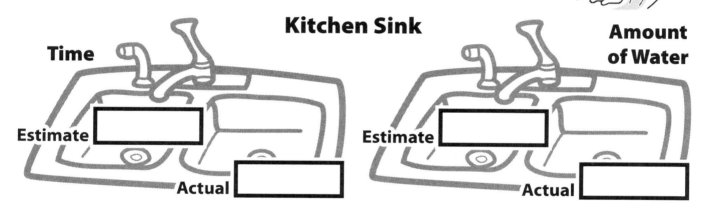

Kitchen Sink

Time

Amount of Water

Estimate

Actual

Estimate

Actual

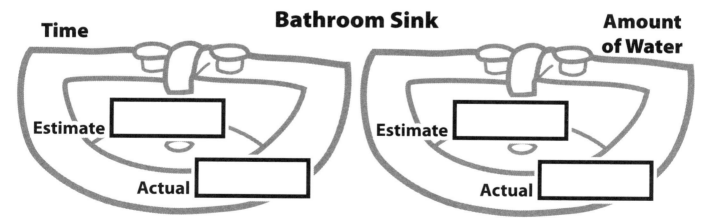

Bathroom Sink

Time

Amount of Water

Estimate

Actual

Estimate

Actual

Is there a difference in the kitchen and bathroom results?
If so, why do you think so?

Down the Drain!

Part Two: School

Trial	Amount of water (L)
1	
2	
3	
Average (median or mean)	

How much water is wasted while waiting for tap water to become warm?

- Estimate how many liters will be lost in one trial:

  [_____ L]

- Put the bucket under a faucet that has not been used for at least 10 minutes.
- Turn on the faucet.
- Turn it off when the water becomes warm.
- Measure the liters of water in the bucket.

Bye-bye

IS IT WARM YET ?

Projections

Assume this is the amount of water wasted daily by each person in your city, state, and country. Calculate the daily and yearly water wasted by people.

	Average wasted	x	Population	=	Amount wasted daily	x 365 =	Amount wasted yearly
City		x		=		x 365 =	
State		x		=		x 365 =	
Country		x		=		x 365 =	

Connecting Learning

Part One: Home

1. How did the times it took your kitchen and bathroom taps to produce warm water compare?

2. What were the shortest and longest times in our class?

3. How did the amount of water from your two taps compare?

4. If the two taps had different results, why do you think so?

5. Whose tap conserved the most water?

Connecting Learning

Part Two: School

1. What was the average amount of water wasted?

2. How close were your estimates to the actual amount of water wasted?

3. What variables might affect how much water is collected?

4. How might you solve the problem of wasting water?

5. What are you wondering now?

Drip Drop ꟼⁱˡⁱ Flop

Topic
Water conservation

Key Question
What are some ways to conserve water around your house?

Learning Goal
Students will respond to pictures of several water-wasting situations by researching and drawing ways to save water around the house.

Guiding Documents
Project 2061 Benchmarks
- *Fresh water, limited in supply, is essential for life and also for most industrial processes. Rivers, lakes, and groundwater can be depleted or polluted, becoming unavailable for life.*
- *The benefits of the earth's resources—such as fresh water, air, soil, and trees—can be reduced by using them wastefully or by deliberately or inadvertently destroying them. The atmosphere and the oceans have a limited capacity to absorb wastes and recycle materials naturally. Cleaning up polluted air, water, or soil or restoring depleted soil, forests, or fishing grounds can be very difficult and costly.*

NRC Standards
- *Resources are things that we get from the living and nonliving environment to meet the needs and wants of a population.*
- *The supply of many resources is limited. If used, resources can be extended through recycling and decreased use.*

NTCM Standard 2000*
- *Represent data using tables and graphs such as line plots, bar graphs, and line graphs*

Math
Graphing
    circle

Science
Environmental science
    natural resources
        water conservation

Integrated Processes
Observing
Comparing and contrasting
Relating

Materials
Crayons or colored pencils
Research resources such as the Internet
Water Conservation rubber band book

Background Information
The average North American uses about 226 liters (60 gallons) of water per day indoors and another 381 liters (100 gallons) outdoors.[1] Compare that to many Sub-Saharan countries, where water use is undesirably low—only 10-20 liters (3-5 gallons) per day per person.[2] In these and other poor countries, people have to contend with water that is not clean and safe to drink; many must walk to a water source and haul the water back in containers. These facts help put into perspective the need for water conservation.

There are many ways that water is wasted around the home. A slowly leaking faucet can waste seven or more gallons a day. Waiting for the water to get hot can waste five gallons. A 15-minute shower uses about 38 gallons. Letting the faucet run while brushing your teeth wastes a gallon or more of water. This activity will help students see ways that water is wasted around the home and help them become more aware of ways they can conserve water.

Water-savers
- take a 3-minute shower instead of a bath
- turn off the faucet while lathering hair
- install a low-flow showerhead
- brush teeth using a cup of water
- install a low-flush toilet
- only run the dishwasher with a full load
- only run the clothes washer with a full load
- keep drinking water in the refrigerator instead of waiting for the faucet to run cold
- fix leaks in faucets and pipes
- water the lawn during the cool part of the day
- wash the car with a bucket or a hose with an automatic shut-off nozzle
- clean driveways and sidewalks with a broom

1   Mayer, P.W. et. al. *Residential End Uses of Water.* American Waterworks Association Research Foundation. Denver, CO. 1999.

2   Cosgrove, William J. and Frank R. Rijsberman. *World Water Vision: Making Water Everybody's Business.* Earthscan Publications Ltd. London, UK. 2000.

Management
To aid students in their research efforts, gather books on water conservation and arrange access to the Internet. Organize computer time for groups of two or three to research together.

Procedure
1. Ask, "How have you used water so far today?" (brushing teeth, washing hands, drinking water, taking a shower, flushing the toilet, etc.) "What room were you in for most of these activities?" (probably the bathroom)
2. Distribute the first activity page. Point out the proportions of water use by location. Instruct students to color and label the circle graph.
3. Tell the class they will be visiting each of these four areas around the house to look for ways to conserve water, starting with the bathroom. Distribute the other student pages. Explain that the first column has pictures of water-wasters. Have students think of ways to save water while doing each task and illustrate their ideas by completing the pictures in the second column.
4. Have students continue drawing water-savers in the kitchen, laundry, and outdoors.
5. Direct students to resources offering more water-saving tips for them to record in the *More water-saving ideas* section of the pages. So students can contribute, listen, and evaluate each other's suggestions, have them conduct research in small groups of two or three. (See *Internet Connections* for suggested websites.)
6. After all the groups have had sufficient research time, bring the class together to discuss their water-saving ideas.
7. Distribute the rubber band book *Water Conservation*. Have students fold it in half width wise with the print on the outside. Have them fold it in half again to make a book.
8. Read the book and talk about any water-saving ideas that weren't previously discussed.

Connecting Learning
1. What are some ways you can save water in the bathroom? …in the kitchen? …in the laundry? …outdoors?
2. In what ways is water being wasted at school? How can we be water-savers at school?
3. What can we do as a class to help the community become better conservers of water?

4. Do you think people in other parts of the world use more or less water than we do in the United States? [less] How can we find some information? [search the Internet, etc.]
5. What are you wondering now?

Extensions
1. Have students research how many gallons of water are used for different activities, like taking a bath, and how much can be saved by changing practices.
2. Encourage students to make posters promoting water conservation. Put them up around the school or community.

Internet Connections
Water Saver Home
http://www.h2ouse.org/
Includes a home tour with facts about water use in each area of the home, suggestions for reducing water use, and a water budget calculator to determine how much water you are using at home.

Water Use It Wisely
http://www.wateruseitwisely.com/
Includes a list of over 100 ways to save water and a home water audit quiz to see how water-wise you are.

Water-saver picture key
Bathroom
   faucet turned off with a cup of water by the sink, half-full tub or short (3-min.) shower
Kitchen
   "all the breakfast, lunch, and dinner dishes," leaky plumbing fixed
Outdoors
   nozzle on hose or a bucket of water, broom to sweep sidewalk
Laundry
   pile of laundry in arms, "I'm washing this big pile of clothes."

* Reprinted with permission from *Principles and Standards for School Mathematics,* 2000 by the National Council of Teachers of Mathematics. All rights reserved.

WATER, PRECIOUS WATER   150   © 2009 AIMS Education Foundation

Drip Drop dıɪꓯ Flop

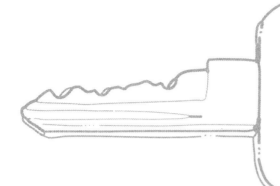

● **Key Question**

What are some ways to conserve water around your house?

Learning Goal

respond to pictures of several water-wasting situations by researching and drawing ways to save water around the house.

Drip Drop Flip Flop

You are going to make a graph that represents water use around the home. The circle is divided into 100 parts. Color

- 21 of those to represent water used in the bathroom,
- 5 to represent water used in the kitchen,
- 9 to represent water used for laundry,
- 59 to represent water used outdoors, and
- 6 to represent leaks and other use.

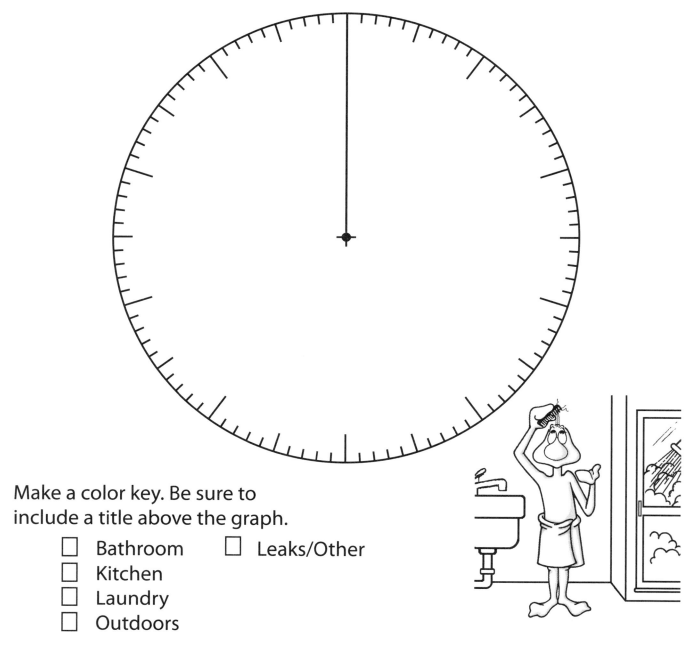

Make a color key. Be sure to include a title above the graph.

- ☐ Bathroom   ☐ Leaks/Other
- ☐ Kitchen
- ☐ Laundry
- ☐ Outdoors

Drip Drop ɟ_ɹ_ıp Flop

Draw a way for each water-waster to turn into a water-saver.
Research and record other ways to save water in the bathroom.

Bathroom

Waster   **Saver**

More water-saving ideas

Drip Drop Flip Flop

Draw a way for each water-waster to turn into a water-saver. Research and record other ways to save water in the kitchen.

Kitchen

Waster   **Saver**

More water-saving ideas

Drip Drop Flip Flop

Draw a way for each water-waster to turn into a water-saver.
Research and record other ways to save water outdoors.

Outdoors

Waster   **Saver**

More water-saving ideas

Drip Drop dᴉlℲ Flop

Draw a way for each water-waster to turn into a water-saver. Research and record other ways to save water in the laundry room.

Laundry

Waster **Saver**

More water-saving ideas

We need to become more aware of ways we can conserve water. There are many simple things that you can do.

There are many ways that water is wasted around the home. A slowly leaking faucet can waste seven or more gallons a day. Waiting for the water to get hot can waste five to 10 gallons. A 15-minute shower uses about 38 gallons. Letting the faucet run while brushing your teeth wastes a gallon or more of water.

Water-savers

- take a 3-minute shower instead of a bath
- turn off the faucet while lathering hair
- install a low-flow showerhead
- brush teeth using a cup of water
- install a low-flush toilet
- only run the dishwasher with a full load
- only run the clothes washer with a full load
- keep drinking water in the refrigerator instead of waiting for the faucet to run cold
- fix leaks in faucets and pipes
- water the lawn during the cool part of the day
- wash the car with a bucket or a hose with an automatic shut-off nozzle
- clean driveways and sidewalks with a broom

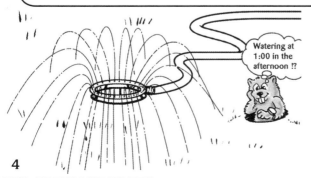

Watering at 1:00 in the afternoon !?

WATER CONSERVATION

drip! drip! drip!

Drip Drop Flip Flop

Connecting Learning

1. What are some ways you can save water in the bathroom? ...in the kitchen? ...in the laundry? ...outdoors?

2. In what ways is water being wasted at school? How can we be water-savers at school?

3. What can we do as a class to help the community become better conservers of water?

4. Do you think people in other parts of the world use more or less water than Americans? How can we find some information?

5. What are you wondering now?

A Little Cup Will Do It!

Topic
Water conservation

Key Question
How much water can be saved in one week when brushing your teeth by using a cup of water instead of letting the water run?

Learning Goals
Students will:
- use 240 mL of water or less for brushing their teeth,
- determine the volume of water they use in a week to brush their teeth, and
- apply their learning to the bigger idea of water conservation.

Guiding Documents
Project 2061 Benchmark
- Fresh water, limited in supply, is essential for life and also for most industrial processes. Rivers, lakes, and groundwater can be depleted or polluted, becoming unavailable for life.

NRC Standards
- Resources are things that we get from the living and nonliving environment to meet the needs and wants of a population.
- The supply of many resources is limited. If used, resources can be extended through recycling and decreased use.

NCTM Standards 2000*
- Understand the effects of multiplying and dividing whole numbers
- Understand such attributes as length, area, weight, volume, and size of angle and select the appropriate type of unit for measuring each attribute
- Select and apply appropriate standard units and tools to measure length, area, volume, weight, time, temperature, and the size of angles

Math
Computation
Measurement
  volume
Graphing

Science
Environmental science
  natural resources
    water conservation

Integrated Processes
Observing
Predicting
Collecting and recording data
Inferring
Applying
Generalizing

Materials
For each student:
  one plastic cup, 9-oz
  one graduated scale

Background Information
Many people allow the tap to run while brushing their teeth. An average faucet has a flow rate of 1.3 gallons per minute. If just over one minute is spent brushing the teeth, this adds up to three gallons (11 liters) per day for two brushings. In one week, the water used to brush teeth while the tap is running totals 21 gallons, almost 80 liters, for each individual. Projected on a yearly basis, a person would use about 1095 gallons or 4145 liters of water while brushing his or her teeth with the tap running. (Realistically, this value would be much lower for school-aged children, who are unlikely to spend more than a minute brushing their teeth.) As the population increases and less usable water is available, we must conserve this valuable resource. This lesson is designed to increase student awareness of ways to conserve water.

Management
1. This activity is to be done at home. Students will monitor their own use of water for one week.
2. Each student will need a 9-oz plastic cup with a graduated scale to use as a measuring cup.
3. It works best to introduce the activity on a Friday and have students begin to gather their data the following Monday morning.

Procedure

1. Discuss water and its importance to living things. Talk about water conservation and have students list some ways to conserve water.
2. Distribute the first student page.
3. Discuss the fact that many people allow the tap to run while brushing their teeth. Brushing twice a day for a little less than one minute each time uses about 7.5 liters of water.
4. Tell students that they are going to be water savers by using water from a cup to brush their teeth rather than letting the faucet run.
5. Give each student a measuring cup. Explain that they are to fill the cup with 240 mL of water each time before brushing. Instead of allowing the faucet to run, they will use the water in the cup to rinse, clean the sink, etc. When they are finished brushing, they will record the amount of water they used by subtracting what is left in the cup from 240.
6. Tell students that they will begin collecting data this coming Monday and that the data will be shared with the class one week later. Be sure that they take home the recording page and the graduated cups.
7. After seven days of gathering data, have the students total the amounts they used in one week and the amount they saved. Pair them up to check each other's computations.
8. Distribute the second student page. Have students use their data to construct a bar graph showing the amount of water used each day of the week. Compare graphs from student to student to see if everyone used about the same amount of water. Discuss any differences.
9. If desired, complete the third student page to determine how much water could be saved by your entire class over one month and one year.

Connecting Learning

1. How much water did you use to brush your teeth each day? Did it change from day to day? Why or why not?
2. How did the amount of water you used brushing your teeth compare to the amount used by others? What might be some reasons for any differences?
3. How much water did you save over one week?
4. Why is water conservation important?
5. What are some other things that you can do to save water?
6. What are you wondering now?

Extensions

1. Older students can be challenged to convert their data into a stacked bar graph for day to day comparisons.
2. Discuss ways to get other people involved in the water conservation campaign.
3. Make posters about water conservation.

* Reprinted with permission from *Principles and Standards for School Mathematics*, 2000 by the National Council of Teachers of Mathematics. All rights reserved.

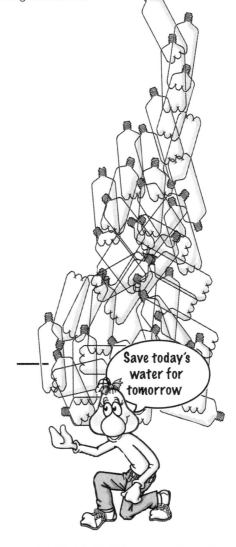

Save today's water for tomorrow

A Little Cup Will Do It!

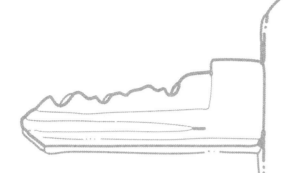

Key Question

How much water can be saved in one week when brushing your teeth by using a cup of water instead of letting the water run?

Learning Goals

Students will:

- use 240 mL of water or less for brushing their teeth,

- determine the volume of water they use in a week to brush their teeth, and

- apply their learning to the bigger idea of water conservation.

A Little Cup Will Do It!

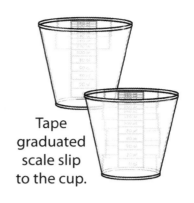

Tape graduated scale slip to the cup.

Graduated Scales
for 9-oz cups

Cover completely with tape.

240	240	240	240	240	240	240	240	240	240
220	220	220	220	220	220	220	220	220	220
200	200	200	200	200	200	200	200	200	200
180	180	180	180	180	180	180	180	180	180
160	160	160	160	160	160	160	160	160	160
140	140	140	140	140	140	140	140	140	140
120	120	120	120	120	120	120	120	120	120
100	100	100	100	100	100	100	100	100	100
80	80	80	80	80	80	80	80	80	80
60	60	60	60	60	60	60	60	60	60
40	40	40	40	40	40	40	40	40	40
20	20	20	20	20	20	20	20	20	20
mL	mL	mL	mL	mL	mL	mL	mL	mL	mL

240	240	240	240	240	240	240	240	240	240
220	220	220	220	220	220	220	220	220	220
200	200	200	200	200	200	200	200	200	200
180	180	180	180	180	180	180	180	180	180
160	160	160	160	160	160	160	160	160	160
140	140	140	140	140	140	140	140	140	140
120	120	120	120	120	120	120	120	120	120
100	100	100	100	100	100	100	100	100	100
80	80	80	80	80	80	80	80	80	80
60	60	60	60	60	60	60	60	60	60
40	40	40	40	40	40	40	40	40	40
20	20	20	20	20	20	20	20	20	20
mL	mL	mL	mL	mL	mL	mL	mL	mL	mL

A Little Cup Will Do It!

When you run the tap while brushing your teeth, you use about one gallon of water! That's equal to 3.8 liters, or 3785 mL. Your goal is to use less water while brushing your teeth by using water from a cup instead of letting the water run. Use this chart to record the water you use each morning and evening while brushing your teeth. At the end of the week, add the two columns on the far right to see how much you used and how much you saved over the whole week.

Day	Water used (A.M.)	+	Water used (P.M.)	=	Total used today	Total with tap running	–	Total used today	=	Total saved
Monday		+		=		7570 mL	–		=	
Tuesday		+		=		7570 mL	–		=	
Wednesday		+		=		7570 mL	–		=	
Thursday		+		=		7570 mL	–		=	
Friday		+		=		7570 mL	–		=	
Saturday		+		=		7570 mL	–		=	
Sunday		+		=		7570 mL	–		=	
						Weekly totals				

Convert the amount of water you saved from milliliters into liters.

$$\underset{\substack{\text{Total mL of} \\ \text{water saved} \\ \text{this week}}}{\underline{}} \text{ mL} \div 1000 \text{ mL} = \underset{\substack{\text{Total number} \\ \text{of liters saved} \\ \text{this week}}}{\underline{}} \text{ L}$$

A Little Cup Will Do It!

Using the data from your table, construct a bar graph showing how much water you used each day of the week. Answer the questions. Use the back of this paper, if needed.

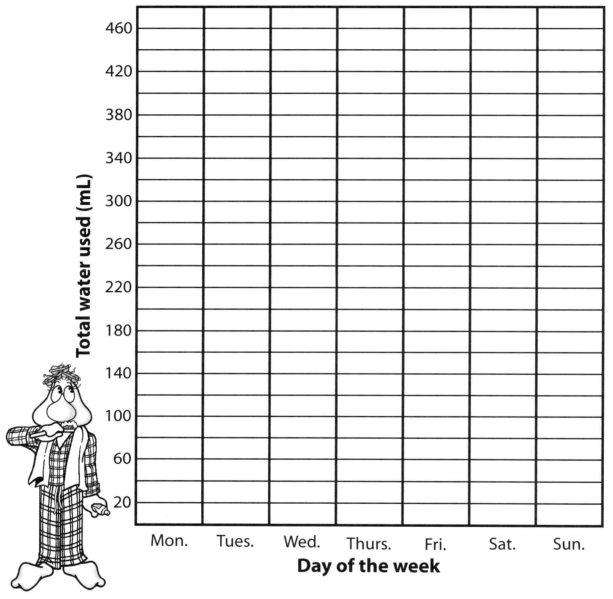

1. Which day did you use the most water?

2. Which day did you use the least water?

3. What changed from day to day that affected how much water you used?

A Little Cup Will Do It!

Calculate how much water would be saved by your class if you all used water from a cup to brush your teeth for an entire year.

Save today's water for tomorrow

Total mL saved for class per day

$$\frac{\text{mL}}{\text{Total number of mL of water saved by all students}}$$

Total liters saved per class per day

$$\frac{\text{Total number of mL saved per class per day}} {} \times \frac{1000 \text{ mL}}{} = \frac{\text{L}}{\text{Total number of liters saved by the whole class}}$$

Total liters saved per class per month

$$\frac{\text{Total number of liters saved per class per day}} {} \times \frac{30 \text{ days}}{\text{Days}} = \frac{\text{L}}{\text{Total number of liters saved per class per month}}$$

Total liters saved per class per year

$$\frac{\text{Total number of liters saved per class per month}} {} \times \frac{12 \text{ months}}{\text{Months}} = \frac{\text{L}}{\text{Total number of liters saved per class per year}}$$

A Little Cup Will Do It!

Connecting Learning

1. How much water did you use to brush your teeth each day? Did it change from day to day? Why or why not?

2. How did the amount of water you used brushing your teeth compare to the amount used by others? What might be some reasons for any differences?

3. How much water did you save over one week?

4. Why is water conservation important?

5. What are some other things that you can do to save water?

6. What are you wondering now?

Water Clock Shower Timer

Topic
Water conservation

Key Question
How much water will you save if you use your water clock to help you take a shorter shower?

Learning Goal
Students will construct and calibrate a water clock to be used as a "shower timer" to help them take shorter showers.

Guiding Documents
Project 2061 Benchmarks
- *Fresh water, limited in supply, is essential for life and also for most industrial processes. Rivers, lakes, and groundwater can be depleted or polluted, becoming unavailable for life.*
- *Organize information in simple tables and graphs and identify relationships they reveal.*

NRC Standards
- *Resources are things that we get from the living and nonliving environment to meet the needs and wants of a population.*
- *Some resources are basic materials, such as air, water, and soil; some are produced from basic resources, such as food, fuel, and building materials; and some resources are nonmaterial, such as quiet places, beauty, security, and safety.*
- *The supply of many resources is limited. If used, resources can be extended through recycling and decreased use.*
- *Use mathematics in all aspects of scientific inquiry.*

NCTM Standards 2000*
- *Select and apply appropriate standard units and tools to measure length, area, volume, weight, time, temperature, and the size of angles*
- *Collect data using observations, surveys, and experiments*

Math
Measurement
  volume
  time
Whole number operations

Science
Environmental science
  natural resources
    water conservation

Integrated Processes
Observing
Collecting and recording data
Comparing and contrasting
Controlling variables

Materials
For each student:
  sticky note
  clear plastic cup, 9 oz
  graduated scale strip
  four jumbo paper clips
  two small paper clips
  one pushpin
  string, 60 cm (24 inches) long
  student pages

For each group:
  graduated cylinder (see *Management 2*)
  transparent tape
  container of water
  one large bowl or bucket
  stopwatch or clock with second hand

Background Information
Water conservation is becoming increasingly important in many urban settings around the United States as drought and other factors place strain on residential water supplies. Of the residential water use that occurs inside the home, almost 20% of that is for showers. (About 2% is for baths.) By shortening the length of showers, significant water savings can be achieved.

According to the energy conservation standards set forth by congress, "the maximum water use allowed for any showerhead manufactured after January 1, 1994, is 2.5 gallons per minute when measured at a flowing water pressure of 80 pounds per square inch."[1] Unless you have very old fixtures, it is likely that your showerhead uses, at most, 2.5 gallons per minute (gpm). In fact, as of 1999, the national average for showerheads was 2.2 gpm[2].

In this lesson, students will make their own calibrated cups to use as shower timers. Each group will have four cups; one with one hole, one with two holes, one with three holes, and one with four holes. Students will observe and record the number of milliliters of water drained from the cups each minute. They should notice that the water leaves the cups more quickly during the first minute because the greater the depth of water, the greater the water pressure, therefore the water is forced through the holes at a faster rate.

As the water level is lowered, there is less water pressure pushing on the remaining water in the cup, slowing the rate of flow. The cup does not completely empty because the minimal amount of water left in the cup covers the holes and surface tension prevents the water from completely dripping out.

1 U.S. Code 6925

2 Mayer, P.W. et. al. *Residential End Uses of Water*. American Water Works Association Research Foundation. Denver, CO. 1999.

Management

1. Each student will need a clear plastic nine-ounce cup to use as a shower timer. Copy the graduated scale strips onto transparency film and cut them apart.
2. Make (or have students make) a graduated cylinder for each group by taping a graduated scale strip to the outside of a clear plastic nine-ounce cup.
3. On a piece of chart paper, make a line with evenly-spaced numbers about 6 cm (2¼") apart. Students will attach sticky notes to show their normal shower times on this line plot.

```
|  1   2   3   4   5   6   7   8   9   10  11  12  |
|                      minutes                     |
```

Procedure

Normal shower data

1. A day before doing this activity, have each student time his or her shower and find out the flow rate, if possible. Offer no further explanation.
2. Give each student a sticky note to place above the appropriate number of minutes on the line plot.
3. Discuss what the line plot shows—range, mode, clusters, gaps, etc.

Water Clock

1. Distribute the first student page and the necessary materials to each student. Have them get into groups of four.
2. Direct students to follow the instructions on the student page to construct their water clocks. Be sure that each person in the group punches a different number of holes in the bottom of the cup.
3. When all students have completed their water clocks, distribute the second student page and discuss the different roles students will have and how they need to be carried out.
4. Distribute the additional materials and have students follow the instructions to determine how long it takes the water to drain from each cup. Be sure they realize that there will always be a small amount of water left in the cup. When the cup hasn't dripped for 10 seconds, they can consider it empty.

5. Once groups have finished and recorded all their data, have groups share with the class. Compare results from group to group and discuss possible reasons for differences.
6. Distribute the final student page. Have students calculate the amount of water they would save in a single shower, a week, and a year if they used one of the water clocks to time their showers.
7. Encourage the students to take the appropriate water clock home and use it in the shower to time their showers. They can tie the full cup of water to the showerhead. When the cup is almost empty or drips slowly, their shower is over!

Connecting Learning

1. How long did it take your cup to empty itself of water?
2. How did this time compare to the time for the other cups in your group?
3. How did the times for cups with the same number of holes compare from group to group?
4. What variables could account for these differences? [Different sized holes, differing degrees of accuracy when reading the volume, etc.]
5. Why will there always be a small amount of water left in the bottom of the cup? [Surface tension prevents the water from dripping out completely.]
6. About how much water do you use in an average shower? [A 10-minute shower uses 25 gallons of water with a 2.5 gpm showerhead, an eight-minute shower uses 20 gallons, and a five-minute shower uses 12.5 gallons.]
7. How much water would be saved if you took a shower that was two minutes shorter? [five gallons]
8. What would your water savings over one year be if you took a shower using your water clock as a timer?
9. Why is it important to conserve water by limiting the amount of water used while showering?
10. What are you wondering now?

Extensions

1. Repeat this experiment with cups that are a different size and shape from the ones used and compare the results. How much impact does the shape of the cup have on the rate at which water drips out?
2. Have the students research the "water clock" as an ancient method of timekeeping.
3. Have the students compute how many milliliters of water they should put into their cups (with less than four holes) to have it empty in three minutes.
4. Brainstorm other methods of personal water conservation that may be practiced or devices that could be made to help save water.

* Reprinted with permission from *Principles and Standards for School Mathematics*, 2000 by the National Council of Teachers of Mathematics. All rights reserved.

Water Clock Shower Timer

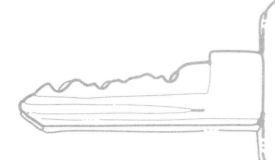

Key Question

How much water will you save if you use your water clock to help you take a shorter shower?

Learning Goal

Students will:

construct and calibrate a water clock to be used as a "shower timer" to help them take shorter showers.

Water Clock Shower Timer

Materials
Clear plastic 9-oz cup
Graduated scale strip
Transparent tape
Pushpin
4 jumbo paper clips
2 small paper clips
60 cm string

Construction

1. Tape your calibrated strip to the outside of the cup and cover it completely with tape.

2. Use a pushpin to punch one, two, three, or four holes in the bottom of your cup. (Each person in your group must choose a different number of holes.)

3. Use a pushpin to make four equally spaced holes around the rim of the cup.

4. Insert a jumbo paper clip into each hole.

5. Loop one small paper clip through the tops of two jumbo paper clips that are next to each other. Repeat with the other two paper clips.

6. Tie the ends of the string to the small paper clips to form a handle. This is your completed water clock.

Water Clock Shower Timer

Your challenge is to find how long it takes for the water to drain from each water clock. There are four roles to fill. Take turns in your group so that everyone gets to do each role once.

Timer: Use a stopwatch or watch with a second hand to time your group. Start the clock when water is poured into the cup. Tell your group each time one minute has passed. When the cup stops dripping, identify how much time has passed.

Measurer: Fill a graduated cylinder with 240 mL of water. When the timer gives you the signal, pour the water into the cup. After each minute has passed, identify the volume of water still in the cup.

Recorder: Record the water level in the cup after each minute using the table on the student page. Record the amount of time it takes for the cup to stop dripping.

Holder: Hold your water clock over the bucket until the water stops dripping. (There will still be water in the cup.)

# of holes	mL after one minute	mL after two minutes	mL after three minutes	mL after four minutes	mL after five minutes	mL after six minutes	mL after seven minutes	mL after eight minutes	mL after nine minutes
1									
2									
3									
4									

Determine the time it took to empty each cup. (A cup is empty when it stops dripping for 10 seconds.) Round your answers to the nearest 30 seconds.

Time to empty each cup

One hole	Two holes	Three holes	Four holes

Water Clock Shower Timer

A regular showerhead uses 2.5 gallons of water per minute. How many gallons of water would you use taking a shower using your water clock as a timer?

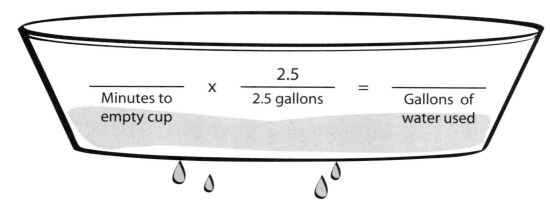

$$\frac{\rule{2cm}{0.4pt}}{\text{Minutes to empty cup}} \times \frac{2.5}{\text{2.5 gallons}} = \frac{\rule{2cm}{0.4pt}}{\text{Gallons of water used}}$$

How long is your average shower? (If you don't know, use 8 minutes.)

My average shower = _____ minutes

Pick a water clock that empties at least one minute faster than your average shower. How many gallons of water would you save per shower using this water clock as your timer?

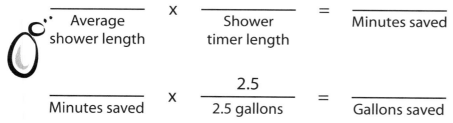

$$\frac{\rule{2cm}{0.4pt}}{\substack{\text{Average}\\\text{shower length}}} \times \frac{\rule{2cm}{0.4pt}}{\substack{\text{Shower}\\\text{timer length}}} = \frac{\rule{2cm}{0.4pt}}{\text{Minutes saved}}$$

$$\frac{\rule{2cm}{0.4pt}}{\text{Minutes saved}} \times \frac{2.5}{\text{2.5 gallons}} = \frac{\rule{2cm}{0.4pt}}{\text{Gallons saved}}$$

How many gallons would you save in one week if you used this water clock as your timer?

$$\frac{\rule{2cm}{0.4pt}}{\text{Gallons saved}} \times \frac{\rule{2cm}{0.4pt}}{\substack{\text{Showers}\\\text{per week}}} = \frac{\rule{2cm}{0.4pt}}{\substack{\text{Gallons saved}\\\text{per week}}}$$

How many gallons would you save in one year if you used this water clock as your timer?

$$\frac{\rule{2cm}{0.4pt}}{\substack{\text{Gallons saved}\\\text{per week}}} \times \frac{52}{\text{52 weeks}} = \frac{\rule{2cm}{0.4pt}}{\substack{\text{Gallons saved}\\\text{per year}}}$$

Connecting Learning

1. How long did it take your cup to empty itself of water?

2. How did this time compare to the time for the other cups in your group?

3. How did the times for cups with the same number of holes compare from group to group?

4. What variables could account for these differences?

5. Why will there always be a small amount of water left in the bottom of the cup?

Connecting Learning

6. About how much water do you use in an average shower?

7. How much water would be saved if you took a shower that was two minutes shorter?

8. What would your water savings over one year be if you took a shower using your water clock as a timer?

9. Why is it important to conserve water by limiting the amount of water used while showering?

10. What are you wondering now?

Water Waste is a Weighty Issue

Topic
Water conservation

Key Questions
1. Would you use as much water each day if you had to carry it into your home by hand?
2. How can you conserve water to help preserve your local water supply?

Learning Goals
Students will:
- learn that the average American uses about 60 gallons a day inside the home,
- experience the difficulty of transporting this amount of water by it carrying it for 30 meters,
- brainstorm ways to reduce the amount of water used in their homes, and
- discuss the positive impact this conservation might have on the local water supply.

Guiding Documents
Project 2061 Benchmark
- *Fresh water, limited in supply, is essential for life and also for most industrial processes. Rivers, lakes, and groundwater can be depleted or polluted, becoming unavailable or unsuitable for life.*

NRC Standard
- *The supply of many resources is limited. If used, resources can be extended through recycling and decreased use.*

Science
Environmental science
    natural resources
        water conservation

Integrated Processes
Observing
Collecting and recording data
Predicting
Generalizing

Materials
For each group of students:
    2-4 plastic gallon containers with handles (see *Management 1*)

    meter sticks or tapes
        (optional: see *Management 3*)
    student pages

Background Information
While water is one of the most abundant substances on Earth, the amount of fresh water available for home use is limited in many areas. In many areas of the country, water from the local watersheds or aquifers is being depleted at a rate that is greater than it is being replenished. Water conservation is an important way to deal with dwindling or limited supplies of water.

According to the American Water Works Association, the average American uses 59.8 gallons of water daily for indoor use. Prior to the widespread availability of indoor plumbing (up until the 1920s, less than one percent of American homes had indoor plumbing), Americans used much less water in their homes. Water was not taken for granted and wasting water was frowned upon, because the water had to be carried by hand from a well or nearby surface-water source (creek, pond, etc.). Since water weighs about eight pounds a gallon, this daily task required a lot of hard work.

It is hoped that this activity will give students an idea of what their ancestors encountered daily as they hauled the water they used. It is also hoped that this experience will help them see the need not to take water for granted and that this will lead to conversations on the importance of conserving water and preserving local water supplies and aquifers.

Management
1. The plastic containers for this activity need to be collected ahead of time. Gallon milk, water, or juice containers with built-in handles work best. Each group will need two to four containers.
2. This activity is designed for students working together in groups of four.
3. In the first part of this activity, students as a group, carry 60 gallons of water a distance of 30 meters to experience the difficulty of this task. The course should be set up as an out-and-back trip of 15 meters each way. You can set it up ahead of time, or students can use meter sticks or tapes to set up the course themselves as part of the activity. The

30 meters represents an average distance that people living before the advent of indoor plumbing would have had to carry their water.

4. The gallon containers need to be filled with water ahead of time.
5. Groups need to devise a method for recording the numbers of gallons carried on the 30-meter course so they know when the 60-gallon goal has been reached. This method can be decided on ahead of time as a class, or left up to each group.
6. If desired, the water carrying part of the activity can be done as a contest to see which group can transport the water in the least amount of time.

Procedure

1. Ask the *Key Questions* and state the *Learning Goals*.
2. Discuss the fact that before the widespread availability of indoor plumbing, most people had to carry all the water they used from a well or surface-water source.
3. Explain the task described on the first student page: Each group will need to transport 60 gallons of water a distance of 30 meters. Discuss how this will be done.
4. After answering any questions, have the groups go outside and complete the task. When all groups are finished, return to the classroom.
5. Discuss the experience and have students answer the question on the first student page.
6. Distribute the second student page. After students have completed it, have a class discussion on what they have learned in this activity.

Connecting Learning

1. If you lived 100 years ago and had to haul your water, where would you build your house?
2. Why do you think cities build up around water sources?
3. People in the United States use more water per person than other people in the world. Why do you think we do? What could we do to reduce this amount?
4. How did this activity help you understand how much water we use?
5. You had four members in your group to carry 60 gallons of water. This represents the water used by just one person. How many gallons would you need to carry for all four people?
6. What are you wondering now?

Extensions

1. Search the Internet for information on local water issues.
2. Research the issue of water conservation. Report your findings.

* Reprinted with permission from *Principles and Standards for School Mathematics, 2000* by the National Council of Teachers of Mathematics. All rights reserved.

Water Waste is a Weighty Issue

Key Questions

1. Would you use as much water each day if you had to carry it into your home by hand?
2. How can you conserve water to help preserve your local water supply?

Learning Goals

Students will:

- learn that the average American uses about 60 gallons a day inside the home,
- experience the difficulty of transporting this amount of water by it carrying it for 30 meters,
- brainstorm ways to reduce the amount of water used in their homes, and
- discuss the positive impact this conservation might have on the local water supply.

Water Waste is a Weighty Issue

The average American uses about 60 gallons of water a day inside the home. This water is used for things like showers, flushing the toilet, and washing clothes.

One hundred years ago, Americans used much less water. At that time, less than one percent of American homes had indoor plumbing. Without plumbing, water had to be carried from a nearby well or surface-water source like a river or pond.

Task: Your group needs to carry a total of 60 gallons of water a distance of 30 meters. Take turns carrying the containers until your group has carried a total of 60 gallons. Carefully keep track of the number of gallons carried as you do this activity.

Would you use as much water each day if you had to carry it by hand into your house? Why or why not?

Water Waste is a Weighty Issue

What are some ways that you might conserve water in your home?

What benefits might conserving water have for your community?

What benefits might conserving water have for your local water supply?

How can you help others in your community see the benefits of conserving water?

Connecting Learning

1. If you lived 100 years ago and had to haul your water, where would you build your house?

2. Why do you think cities build up around water sources?

3. People in the United States use more water per person than other people in the world. Why do you think we do? What could we do to reduce this amount?

4. How did this activity help you understand how much water we use?

Connecting Learning

5. You had four members in your group to carry 60 gallons of water. This represents the water used by just one person. How many gallons would you need to carry for all four people?

6. What are you wondering now?

What if I had to carry all that water from...

HOME 5 MILES

Topic
Water distribution

Key Question
What is the least expensive way to design a water system that would deliver 1000 gallons of water to each city?

Learning Goals
Students will:
- learn the importance of sharing a limited natural resource, in this case water, throughout a given area;
- use higher-level thinking skills to solve the water distribution problem by designing an efficient water system;
- discuss and analyze possible solutions and realize that natural resources such as water must be shared by all; and
- recognize the effect of human presence and activity on our water sources.

Guiding Documents
Project 2061 Benchmarks
- *Fresh water, limited in supply, is essential for life and also for most industrial processes. Rivers, lakes, and groundwater can be depleted or polluted, becoming unavailable or unsuitable for life.*
- *Scientific laws, engineering principles, properties of materials, and construction techniques must be taken into account in designing engineering solutions to problems. Other factors, such as cost, safety, appearance, environmental impact, and what will happen if the solution fails also must be considered.*
- *Technologies often have drawbacks as well as benefits. A technology that helps some people or organisms may hurt others—either deliberately (as weapons can) or inadvertently (as pesticides can). When harm occurs or seems likely, choices have to be made or new solutions found.*
- *Because of their ability to invent tools and processes, people have an enormous effect on the lives of other living things.*

NRC Standards
- *Communicate investigations and explanations.*
- *Technological designs have constraints. Some constraints are unavoidable, for example, properties of materials, or effects of weather and friction; other*

constraints limit choices in the design, for example, environmental protection, human safety, and aesthetics.
- *Technological solutions have intended benefits and unintended consequences. Some consequences can be predicted others cannot.*
- *Human activities also can induce hazards through resource acquisition, urban growth, land-use decisions, and waste disposal. Such activities can accelerate many natural changes.*

Math
Problem solving
Computation

Science
Enviromental science
    natural resources
        water use/distribution

Integrated Processes
Observing
Collecting and recording data
Comparing and contrasting
Generalizing

Materials
For each group:
    crayons or colored pencils
    glue
    tape
    scissors
    student pages

Key Vocabulary
Aqueduct: a canal or water distribution system
Dam: a structure that allows water to be stored in a reservoir or lake
Desalinization: a process of removing salt from ocean water
Ground water: the water table; an underground water supply
Pipeline: pipes that carry water, either underground or above ground
Pump station: a station that is necessary to pump water around a city or over elevated terrain
Reservoir: a storage area of water
Water purification plant: a plant that purifies water
Well: a means of obtaining ground water

Background Information

There are various sources of water that we use for our fresh water needs. Water is obtained from ground water sources, lakes, rivers, and desalinization plants. Each of these sources has different advantages and disadvantages. Some processes and methods used to obtain water have more impact on the environment and the water sources than others. Ground water supplies can be polluted or depleted by overuse. Dams create changes to the landscape and can affect plant and animal life in the area. Desalinization plants are costly and can affect the ocean areas where they are located.

Management

1. This is a simulation and the rules should be adjusted to meet your class's needs.
2. Students should work together in groups of two to four.

Procedure

Part One

1. Hand out one set of island map pages to each group and have students tape them together. Discuss the geographical features with the students. Instruct students to color the surface water sources blue and color the underground water purple.
2. Ask the students to describe each city and discuss possible ways to supply its water. List these ideas on the board. Use the *Key Vocabulary* page to introduce the terms used in this activity.
3. Hand out the *Construction Rules* and *Water Works Sheet* to each group, and as a class discuss the benefits, drawbacks, and uses of each item.
4. Explain that each source or distribution method has both a point value related to cost and a maximum amount of water it can provide.
5. Tell students that the object of this activity is to plan one or more solutions to the water needs of both cities. The goal is to design a system that will bring 1000 gallons of water to any point along the city limits line using the least amount of points.
6. The *Construction Rules* table includes the rules to be followed. Remind students to observe these rules.
7. Have each team determine their best solution and cut and paste it in place on the island map.
8. Instruct students to write an explanation of their distribution plan. Using their *Water System Cost Sheet,* have them total the cost of their systems and record the number of points it will take.
9. Allow time for each group to share its final plan with the class. Have students record each group's results on the *Group Cost Comparison Sheet.*
10. Have two groups meet to share and compare their water maps and distribution plans. Have them record the differences between their plans.

Part Two

1. Repeat the activity, giving each group a different *Fate Card*. Have them adjust their water systems accordingly.
2. Discuss the potential impacts of human presence on our water sources.

Connecting Learning

Part One

1. How were the water needs of each city different?
2. Explain the process you used to make your first plan.
3. Did your first plan change? Why or why not?
4. What did you learn about how geography affects the distribution of water?
5. What is the danger of having a single source of water?
6. What would be the effect of a third city being added to Water Island? Would the location matter? Why or why not?

Part Two

1. How did the fate card change your water distribution system?
2. Can you think of any examples of similar things happening in our area or in other parts of the country? What are they?
3. Which fate cards were results of human presence or activity? Which were the results of natural forces?
4. What kinds of things can we do to reduce the negative impact we have on our fresh water supplies?
5. What are you wondering now?

Extensions

1. Make an efficient water network between the two cities.
2. Repeat the activity and assign varying populations and water demands to each city. Each person would need 100 gallons of water.
3. Challenge students to add a third city to Water Island in the location where it will have the least impact on the water supply.

Curriculum Correlation

Social Science

a. Research a real island to find out how the island's water is supplied.
b. Add coordinates to the island map. Select a location for a hidden treasure and give directions so the students can find the treasure.

WATER ISLAND

Key Question

What is the least expensive way to design a water system that would deliver 1000 gallons of water to each city?

Learning Goals

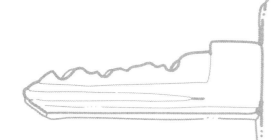

Students will:

- learn the importance of sharing a limited natural resource, in this case water, throughout a given area;
- use higher level thinking skills to solve the water distribution problem by designing an efficient water system;
- discuss and analyze possible solutions and realize that natural resources such as water must be shared by all; and
- recognize the effect of human presence and activity on our water sources.

Key Vocabulary

Aqueduct: a canal or water distribution system

Dam: a structure that allows water to be stored in a reservoir or lake

Desalinization: a process of removing salt from ocean water

Ground water: the water table; an underground water supply

Pipeline: pipes that carry water, either underground or above ground

Pump station: a station that is necessary to pump water around a city or over elevated terrain

Reservoir: a storage area of water

Water purification plant: a plant that purifies water

Well: a means of obtaining ground water

BLUE OCEAN

CITY A

RED RIVER
(100 gallons per day)

ground water
five water drops = water for one well

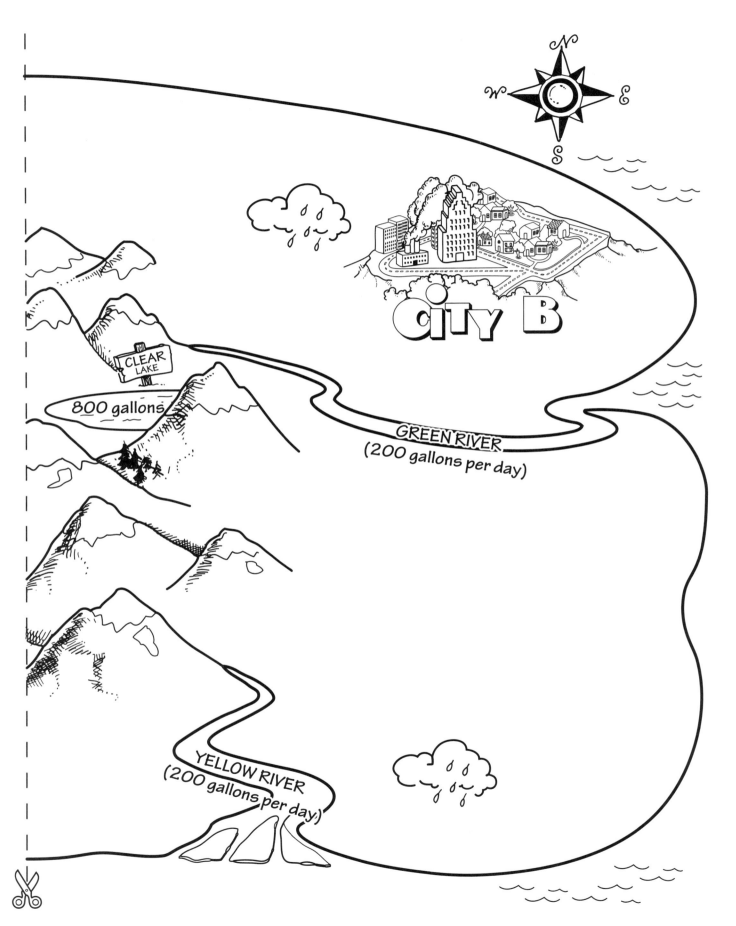

CLEAR
LAKE

800 gallons

CITY B

GREEN RIVER
(200 gallons per day)

YELLOW RIVER
(200 gallons per day)

WATER ISLAND
WATER WORKS

RESERVOIR DAM 100

WATER PURIFICATION PLANT 60 — STORAGE

PUMP STATION 30

PUMP STATION 30

Well 20

Well 20 — DESALINIZATION PLANT 80 PTS. 400 GAL.

WATER ISLAND
Construction Rules

Item	Special notes	Cost in points	Gallons of water per day
One well	Equal to five raindrops of ground water	20 points	100 gallons
One dam with reservoir	All rivers require dams	100 points	—
One pump station	Needed to move water anywhere One pump station per 3 aqueducts/pipes	30 points	—
One water purificaiton plant	Water storage—one per city	60 points	—
One desalinization plant		300 points	400 gallons
One section of aqueduct/pipeline		10 points	—

Water System Cost Sheet

City A				City B		
Item	Cost in points	Total gallons water		Item	Cost in points	Total gallons water
1.				1.		
2.				2.		
3.				3.		
4.				4.		
5.				5.		
6.				6.		
7.				7.		
8.				8.		
Total:				Total:		

WATER ISLAND

Group Cost Comparison Sheet

Group	Total cost (points) City A	+	Total cost (points) City B	=	Total island water cost
1.		+		=	
2.		+		=	
3.		+		=	
4.		+		=	
5.		+		=	
6.		+		=	
7.		+		=	
8.		+		=	

Put a star in front of the number of your group and complete the table with the data from the rest of the class.

1. Which group had the lowest cost in points for City A? ...for City B?

2. How did any other group's cost differ from your group? Why?

For the last three years, low rainfall In the southwestern region has *reduced* the ground water supply by 20 drops.

Adjust your water system.

FATE CARD

The salt content in the water has increased! The salt needs to be removed from the water and the cost for water purification is doubled.

Adjust your water system.

FATE CARD

Low snowpack for the last two years has resulted in a reduction in the amount of water in the lake by 200 gallons.

Adjust your water system.

FATE CARD

A parasite has multiplied in the Green River. No water can be used for two years.

Adjust your water system.

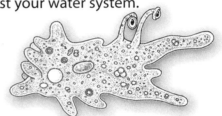

FATE CARD

Citizens of City A have polluted their ground water by using insecticides. No ground water in the area can be used.

Adjust your water system.

FATE CARD

City A's population has increased. You will need to increase the water supply to 1500 gallons per day.

Adjust your water system.

FATE CARD

A factory has dumped its sludge into the Yellow River No river water can be used!

Adjust your water system.

FATE CARD

Flooding has washed out the aqueducts and rerouted the Green River 5 cm south on your map.

Rebuild your aqueduct system.

FATE CARD

WATER ISLAND

Connecting Learning

Part One

1. How were the water needs of each city different?

2. Explain the process you used to make your first plan.

3. Did your first plan change? Why or why not?

4. What did you learn about how geography affects the distribution of water?

5. What is the danger of having a single source of water?

6. What would be the effect of a third city being added to Water Island? Would the location matter? Why or why not?

WATER ISLAND

Connecting Learning

Part Two

1. How did the fate card change your water distribution system?

2. Can you think of any examples of similar things happening in our area or in other parts of the country? What are they?

3. Which fate cards were results of human presence or activity? Which were the results of natural forces?

4. What kinds of things can we do to reduce the negative impact we have on our fresh water supplies?

5. What are you wondering now?

Topic
Water treatment: settling

Key Question
How can settling be used to clean water that is dirty?

Learning Goal
Students will mix sand, oil, and water and see how the mixture settles.

Guiding Documents
Project 2061 Benchmarks
- *Fresh water, limited in supply, is essential for life and also for most industrial processes. Rivers, lakes, and groundwater can be depleted or polluted, becoming unavailable or unsuitable for life.*
- *Measure and mix dry and liquid materials (in the kitchen, garage, or laboratory) in prescribed amounts, exercising reasonable safety.*

NRC Standard
- *Natural environments may contain substances (for example, radon and lead) that are harmful to human beings. Maintaining environmental health involves establishing or monitoring quality standards related to use of soil, water, and air.*

NCTM Standard 2000*
- *Select and apply appropriate standard units and tools to measure length, area, volume, weight, time, temperature, and the size of angles*

Math
Measurement
    mass
    volume

Science
Physical science
    mixtures
Environmental science
    water treatment
       settling

Integrated Processes
Observing
Predicting
Comparing and contrasting
Interpreting
Generalizing

Materials
For each group:
    plastic cup, 9 oz
    graduated strip
    clear testing jar with a lid and straight sides
    vegetable oil
    clean sand
    balance
    gram masses
    food coloring
    centimeter ruler
    water

Background Information
A mixture is a combination of substances in which the parts can be separated. Settling is one method of separating a mixture of liquids and solids. When a mixture of liquids and solids settles, some material sinks to the bottom and some floats to the top. Layering will occur due to the differing densities of the materials. The bottom layers will be the most dense and the top layers the least dense. In this activity, the sand will settle to the bottom and the oil will float to the top. Settling, usually called sedimentation, is a key step in the water treatment process.

Management
1. Copy the graduated strips onto transparency film. Tape one strip to the outside of each 9-oz cup to be used as a measuring device.
2. Each group will need a clear jar with a screw on lid and straight sides. Peanut butter jars, jam jars, or olive jars all work well.
3. Sand can be purchased from building supply or home improvement stores. If clean sand is not available, rinse the sand and pour off the dirty water until the water runs clear.
4. Divide the class into groups of three or four.

Procedure
1. Explain that it is important for people's health that water is safe to drink. Ask students to give some examples of safe drinking water such as treated water coming through faucets and bottled water. Present the *Key Question:* "How can settling be used to clean water that is dirty?"
2. Show students the oil, water, and sand. Tell them that they will observe what happens when they make a mixture of these three substances in a jar.

3. Distribute the student pages and have students draw a picture in the first jar of how they think it will look after pouring in the three items—first the water, then the oil, then the sand.

4. Arrange the students into their groups, and distribute the materials.

5. Instruct groups to measure 100 mL of water using the calibrated measuring cup and pour it into their testing jars. Have them add two drops of food coloring to the water and swirl it to mix.

6. Instruct groups to measure 100 mL of vegetable oil using the calibrated measuring cup and pour it into the testing jar.

7. Have students use the balance to measure 50 grams of clean sand and add it to the testing jar.

8. When all three items have been added, have students draw a picture on the activity page showing what the mixture looks like after pouring but before shaking.

9. Using a centimeter ruler, have each group measure and record the height of the water in the jar.

10. Have students predict what the jar will look like after shaking and draw a picture in the appropriate jar on the activity page showing their predictions.

11. Be sure that the lids are screwed on to the containers tightly. Have each group shake their jar vigorously for at least 10 seconds and draw a picture showing what the mixture looks like immediately after shaking.

12. As they are waiting for the mixture to settle, have them illustrate their predictions of how the jars will look after 10 minutes.

13. Allow the mixtures to settle for about 10 minutes. Have students complete the activity page by drawing what the jar looks like after settling.

14. Discuss how settling, or sedimentation, would be useful in water treatment.

Connecting Learning

1. How did you think the oil, water, and sand would distribute themselves in the jar?

2. How did this compare to the way they actually looked?

3. What happened to the materials when you shook the jar? Why?

4. Where in the real world would you find conditions similar to the jar after it was shaken? [in the ocean along the coasts, in rivers, etc.]

5. What did the jar look like after you let it settle for 10 minutes? Why?

6. What happened to the height of the water before and after shaking?

7. Why would this method be used in the process of cleaning our drinking water?

8. What are you wondering now?

Extensions

1. Have students experiment with other items to create different mixtures.

2. Give students a dirty water mixture for them to clean by applying their newly gained knowledge on settling.

That Settles It

Key Question

How can settling be used to clean water that is dirty?

Learning Goal

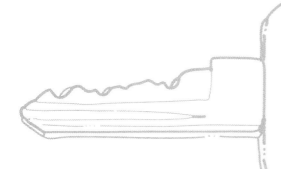

Students will:

mix sand, oil, and water and see how the mixture settles.

THAT SETTLES IT

Draw a series of pictures showing what happens when oil, water, and sand are mixed and then allowed to settle. In each case, draw your prediction and the actual result.

10 minutes after shaking

Prediction

Actual

Immediately after shaking

Prediction

Actual

After pouring the water, oil, and sand

Prediction

Actual

THAT SETTLES IT

Measure the height of the water before you shake the container and 10 minutes after shaking the container. Record the values below.

Before shaking: _____ cm

After settling: _____ cm

Why do you think the materials in the jar arranged themselves the way they did?

Why would settling be useful in water treatment?

That Settles It

Connecting Learning

1. What was your prediction for how the oil, water, and sand would distribute themselves in the jar?

2. How did this compare to the way they actually looked?

3. What happened to the materials when you shook the jar? Why?

4. Where in the real world would you find conditions similar to the jar after it was shaken?

Connecting Learning

5. What did the jar look like after you let it settle for 10 minutes? Why?

6. What happened to the height of the water before and after shaking?

7. Why would this method be used in the process of cleaning our drinking water?

8. What are you wondering now?

Help Save the Birds!

Topic
Water treatment: filtration

Key Question
How can you make muddy water cleaner so that it is clear enough for birds to bathe in?

Learning Goal
Students will use the process of filtration, an important step in water purification, to devise a system to filter dirty water.

Guiding Documents
Project 2061 Benchmark
- *Changes in an organism's habitat are sometimes beneficial to it and sometimes harmful.*

NRC Standards
- *Humans depend on their natural and constructed environments. Humans change environments in ways that can be either beneficial or detrimental for themselves and other organisms.*
- *Identify a simple problem.*
- *Propose a solution.*
- *Implementing proposed solutions.*
- *Evaluate a product or design.*
- *Communicate a problem, design, and solution.*
- *The supply of many resources is limited. If used, resources can be extended through recycling and decreased use.*
- *Changes in environments can be natural or influenced by humans. Some changes are good, some are bad, and some are neither good nor bad. Pollution is a change in the environment that can influence the health, survival, or activities of organisms, including humans.*

Math
Measurement
 volume

Science
Environmental science
 water treatment
  filtration
Physical science
 chemistry
  separation of mixtures

Integrated Processes
Observing
Comparing and contrasting
Collecting and recording data

Materials
For the class:
 2 liters of pre-made muddy water (see *Management 2*)
 2-3 plastic or Styrofoam cups
 clear plastic cups, 9 oz
 graduated scale (see *Management 3*)
 aluminum foil
 paper towels
 clean, fine sand
 optional: clean gravel, small aquarium rocks, charcoal, cotton, etc.

Background Information
   Filtration is one method for separating a mixture. Many water treatment plants have three basic steps in the process of purifying the water: 1) coagulation and settling, 2) filtration, 3) dissection.
   In filtration, water is passed through a bed of sand usually about 2 ½ feet deep on top of a bed of gravel about 1 foot deep. The water is then treated by adding chlorine to kill disease-carrying organisms.

Management
1. You may organize this lesson as an open-ended, divergent activity or as a teacher-directed lesson.
   a. Open-ended lesson: Allow the students to select their materials from a "supply table" that has all the suggested materials.
   b. Teacher-directed lesson: Have students all make the same filtering device, step-by-step, from a given set of materials.
2. When making the muddy water, be sure to include bits of leaves, sticks, sand, dirt, etc. Each group of four students will need 240 mL of dirty water.

3. Copy the graduated scale onto transparency film. Cut out each scale and tape to the outside of the 9-oz plastic cups. This can be done beforehand or students can do it in preparation for the activity.

Hints for making a filtering device:
a. Poke holes in a Stryofoam cup and place a paper towel in it to keep the sand and gravel from falling through.
b. Make a funnel out of aluminum foil and line it with a paper towel and fill it with sand and gravel.

Procedure
1. Set the stage with the students by asking them to imagine the following situation in a local park:

   "Your town has just experienced a severe rainstorm with some minor flooding. The local park is very soggy and filled with mud holes. The birds used to bathe themselves in the clean water areas in the park but now they are filled with muddy water. How can we clean the water for the birds?"
2. Ask the *Key Question* and state the *Learning Goal*.
3. Show students the sign that was posted in the park.
4. Hold up a cupful of muddy water and ask students to design a method for cleaning the muddy water.
5. Allow the students to select the materials from a class "supply table" or distribute materials, as listed, to each group.
6. Hand out the student page to each group. Discuss the page.
7. Have each group construct a filtration system.
8. Have the students measure 240 mL of dirty water and pour it through their filtration system.
9. Collect the filtered water in a separate cup and have students measure the amount reclaimed. Instruct them to record the amount and complete the student page.

Connecting Learning
1. Explain your procedures for cleaning the water.
2. How much water were you able to reclaim?
3. Were your results similar to other groups? Explain.
4. How much water was not reclaimed?
5. What were the "costs" of having to clean the water? [Not only did we have to use materials such as paper towels, sand, and aluminum foil, but it cost us some of the water, too.]
6. Were you satisfied with the cleanliness of the water? Explain your answer.
7. In what cases would this type of filtration be "good enough"?
8. Why would water be considered a renewable resource?
9. Why would water be considered a nonrenewable resource?
10. What are you wondering now?

Extensions
1. Collect all the clean water samples from each group and put them outside in a shallow pan for the birds to bathe in around the school. Try to locate the "bird bath" near the classroom so you can check on it during the day.
2. Try placing 4-5 drops of food coloring in the muddy water to represent contaminants in the water system. Observe what happens to the cleansed water with the food coloring in it. Does it filter out? Discuss water pollution.
3. Evaporate a small sample of each group's cleansed water to check for other particles, minerals, etc., that may have been left after filtering.

* Reprinted with permission from *Principles and Standards for School Mathematics,* 2000 by the National Council of Teachers of Mathematics. All rights reserved.

Help Save the Birds!

Key Question

How can you make muddy water cleaner so that it is clear enough for birds to bathe in?

Learning Goal

use the process of filtration, an important step in water purification, to devise a system to filter dirty water.

Help Save the Birds!

Graduated Scales
for 9-oz cups

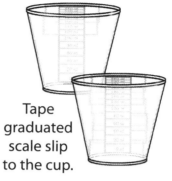

Tape graduated scale slip to the cup.

Cover completely with tape.

240	240	240	240	240	240	240	240	240	240
220	220	220	220	220	220	220	220	220	220
200	200	200	200	200	200	200	200	200	200
180	180	180	180	180	180	180	180	180	180
160	160	160	160	160	160	160	160	160	160
140	140	140	140	140	140	140	140	140	140
120	120	120	120	120	120	120	120	120	120
100	100	100	100	100	100	100	100	100	100
80	80	80	80	80	80	80	80	80	80
60	60	60	60	60	60	60	60	60	60
40	40	40	40	40	40	40	40	40	40
20	20	20	20	20	20	20	20	20	20
mL	mL	mL	mL	mL	mL	mL	mL	mL	mL

240	240	240	240	240	240	240	240	240	240
220	220	220	220	220	220	220	220	220	220
200	200	200	200	200	200	200	200	200	200
180	180	180	180	180	180	180	180	180	180
160	160	160	160	160	160	160	160	160	160
140	140	140	140	140	140	140	140	140	140
120	120	120	120	120	120	120	120	120	120
100	100	100	100	100	100	100	100	100	100
80	80	80	80	80	80	80	80	80	80
60	60	60	60	60	60	60	60	60	60
40	40	40	40	40	40	40	40	40	40
20	20	20	20	20	20	20	20	20	20
mL	mL	mL	mL	mL	mL	mL	mL	mL	mL

Attention All Bird Lovers:

The water in the park is very muddy since the last rainstorm. The birds need clean, clear water in which to wash their feathers.

Please Help Us...

We need someone to find a way to make the water as clean and clear as possible.

Help stamp out "Dirty Birds!"

Help Save the Birds!

Members of the "Help Save the Birds" committee:

○ _____ ○ _____ ○ _____
○ _____ ○ _____ ○ _____

Draw a picture of your group's filtration system. Label each item in your system.

List your group's procedures:

Amount of water given to clean = _____ mL

Amount of water your group cleaned = _____ mL

Difference = Amount of water trapped in your filter _____ mL

Help Save the Birds!

Connecting Learning

1. Explain your procedures for cleaning the water.

2. How much water were you able to reclaim?

3. Were your results similar to other groups? Explain.

4. How much water was not reclaimed?

5. What were the "costs" of having to clean the water?

Help Save the Birds!

Connecting Learning

6. Were you satisfied with the cleanliness of the water? Explain your answer.

7. In what cases would this type of filtration be "good enough"?

8. Why would water be considered a renewable resource?

9. Why would water be considered a nonrenewable resource?

10. What are you wondering now?

MiNi Water Treatment Simulation

Topic
Water treatment

Key Question
How is water from the natural water cycle purified for home use?

Learning Goal
Students will simulate the steps in the typical water treatment process.

Guiding Documents
Project 2061 Benchmark
- *Geometric figures, number sequences, graphs, diagrams, sketches, number lines, maps, and stories can be used to represent objects, events, and processes in the real world, although such representations can never be exact in every detail.*

NRC Standards
- *Technology influences society through its products and processes. Technology influences the quality of life and the ways people act and interact. Technological changes are often accompanied by social, political, and economic changes that can be beneficial or detrimental to individuals and to society. Social needs, attitudes, and values influence the direction of technological development.*
- *When an area becomes overpopulated, the environment will become degraded due to the increased use of resources.*

Science
Environmental science
   water treatment

Integrated Processes
Observing
Comparing and contrasting
Applying
Generalizing

Materials
For each group:
   water
   dirty water (see *Management 1)*
   two clear plastic cups, 10-12 oz
   Styrofoam cups
   powdered alum (see *Management 2)*
   1/8 teaspoon measure
   coffee filters (see *Management 3)*
   clean sand
   clean gravel
   yellow food coloring (see *Management 6)*
   eyedropper
   first student page

For each student:
   The Water Treatment Process rubber band book
   #19 rubber band
   second and third student pages
   scissors
   tape

Background Information
A water company must go through several steps to ensure safe and pure drinking water for a community. The water that is processed comes from the natural water cycle and has usually been transferred and stored in a reservoir before processing.

The following steps are found in a typical water treatment plant:

Aeration: Water is sprayed into the air to release any trapped gases and to absorb additional oxygen.

Coagulation: To remove dirt suspended in the water, powdered alum is dissolved in the water, forming tiny sticky particles called floc that attach to the dirt particles. The combined weight of the dirt and the alum particles (floc) becomes heavy enough to sink to the bottom during sedimentation.

Sedimentation: The heavy particles (floc) settle to the bottom and the clear water above the particles is skimmed from the top and is ready for filtration.

Filtration: The clear water passes through layers of sand, gravel, and charcoal to remove small particles.

Chlorination: A small amount of chlorine gas is added to kill any bacteria or microorganisms that may be in the water.

Management
1. Add one-half teaspoon of dirt for each one cup of water to make the dirty water mixture used in this activity. Each group needs a 10- to 12-ounce cup of dirty water.
2. This activity may be done without adding the alum. However, if you do add the alum, it produces much clearer water. Ammonium alum is quite inexpensive and may be purchased at any drug store. (This is not the alum that is found in grocery stores.) If it

is not in stock, ask the manager to order it for you. Druggists can usually get it in a day or two. The alum creates "floc" that may take 10-15 minutes to settle to the bottom. It is well worth the effort.

3. Cut coffee filters into quarters so that they will easily line the Styrofoam cups.
4. Clean sand may be purchased from any hardware store or home-supply store. If you use sand from the playground, be sure to rinse it well first to remove any dirt.
5. Aquarium gravel can be purchased from a pet store and works well.
6. Make a mixture of two drops of yellow food coloring in a cup of water to simulate the chlorine that is added to water for purification.
7. Gather the materials for the simulation and place them on a table or other central location where groups can come to collect what they need.
8. Copy the third student page on card stock.

Procedure

1. Ask the *Key Question* and state the *Learning Goal.*
2. Discuss water purification. Inform students that in order to have safe and pure drinking water, the local supply must go through several steps in a treatment process.
3. Distribute the rubber band book to each student. Read through it as a class and discuss the steps in the water treatment process.
4. Divide students into groups and distribute one clear plastic cup filled with dirty water (see *Management 1*). Tell them that water that has come through the natural water cycle might not be as dirty as this sample.
5. Hand out a copy of the first student page to each group and a copy of the second student page to each student. Go over the instructions for each step of the water treatment simulation and show students where they can collect the materials that they will need.
6. Ask one person from each group to gather the necessary materials.
7. Have students go through the water filtration process and record their observations on the second student page. In step four, students may make a filter using multiple Styrofoam cups that each contain a separate material through which to filter the water. In step five, students should add one or two drops of **simulated** chlorine bleach to the final water sample. **CAUTION: Do not use real bleach or drink the water!**
8. When groups have finished the simulation, distribute the third student page. Have students cut out each of the pictures below the water treamtent plant and arrange them in the proper order to show the sequence of water treatment.

9. Tell students to tape all of the pictures together in order to form a long strip and pass it through the slits on the water treatment plant to show the steps in the water treatment process.

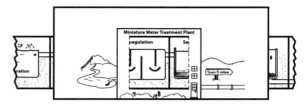

Connecting Learning

1. What are the steps of water treatment that we simulated?
2. What do you think would have happened if the steps were done in a different order?
3. How was our simulation like the real process of water treatment? How was it different?
4. Which of the processes do you think had the greatest impact on cleaning the water?
5. How do the cleaned samples compare from group to group? What might be reasons for any differences?
6. Why aren't you allowed to drink the water in this simulation?
7. How could we find out how our local water is purified?
8. What are you wondering now?

Extensions

1. Investigate local water treatment procedures by writing letters, using the Internet, or interviewing local water officials.
2. Go on a field trip to the local water treatment facility.

Solutions

The correct order for the strip on the third student page is as follows.

1. Water from lakes, streams and reservoirs
2. Aeration
3. Alum
4. Coagulation
5. Sedimentation
6. Filtration
7. Chlorination
8. Storage
9 To the city homes

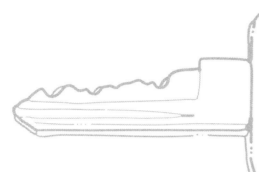Water Treatment Simulation

Key Question

How is water from the natural water cycle purified for home use?

Learning Goal

simulate the steps in the typical water treatment process.

Finally, a small amount of chlorine gas is added to the water. This kills any bacteria or microorganisms. This is called **chlorination**.

CHLORINATION

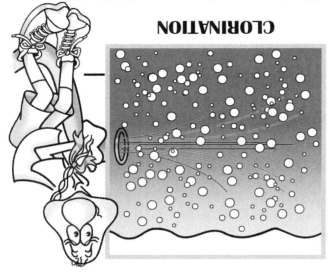

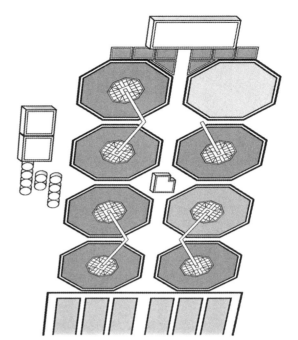

Before water goes to houses, it goes through a water treatment plant. The treatment plant makes sure water is safe for people to use.

Not every water treatment plant uses all of these steps in the exact same way. But all the water from the tap has gone through a water treatment process.

THE WATER TREATMENT PROCESS

How do they make sure it's safe?

SEDIMENTATION

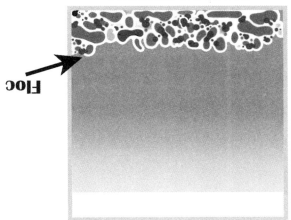

Floc

The floc and the dirt sink to the bottom. This is called **sedimentation**. The clear water above the particles is skimmed from the top. Now it is ready for filtration.

Next comes **coagulation**. Powdered alum is dissolved in the water. It forms tiny sticky particles called floc that attach to the dirt particles.

COAGULATION

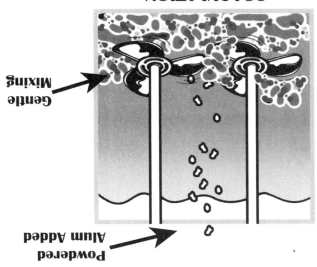

Gentle Mixing

Powdered Alum Added

In **filtration**, the clear water passes through layers of sand, gravel, and charcoal to remove small particles.

Hmm... which would you rather drink?

Unfiltered Filtered

FILTRATION

Water Plant

There are several steps to treating water. The first step is **aeration**. Water is sprayed into the air. This releases any trapped gases and lets it absorb additional oxygen.

Mini Water Treatment Simulation

Step One: Aeration
- Fill one clear pastic cup with dirty water.
- Pour the water back and forth between the two large plastic cups.
- Record your observations.

1/8 teaspoon alum →

Step Two: Coagulation
- Sprinkle 1/8 teaspoon of alum on top of the water.
- Record your observations.

Step Three: Sedimentation
- Let the water sit for 10-15 minutes.
- After the floc has settled to the bottom of the cup, carefully pour the water into the prepared filter. Leave the floc in the bottom of the cup.
- Record your observations.

Step Four: Filtration
- While the water is settling, make a filter from a Styrofoam cup. Poke 10 holes in the bottom of the cup with a sharpened pencil.
- Line the cup with a piece of coffee filter.
    - Fill the cup with layers of clean sand and gravel.
    - Pour the water through the filter and collect it in the other large plastic cup.
        - For an even better filter, make multiple filter cups and stack them on top of each other.
        - Record your observations.

Layers of clean sand and gravel

Coffee filter for liner

Step Five: Disinfection
- Put one or two drops of simulated bleach in the water. **DO NOT DRINK THE WATER!**
- Record your observations.

MINI Water Treatment Simulation

Record your observations for each step of the water treatment simulation. Use words and pictures.

Step One: Aeration

Step Two: Coagulation

Step Three: Sedimentation

Step Four: Filtration

Step Five: Disinfection

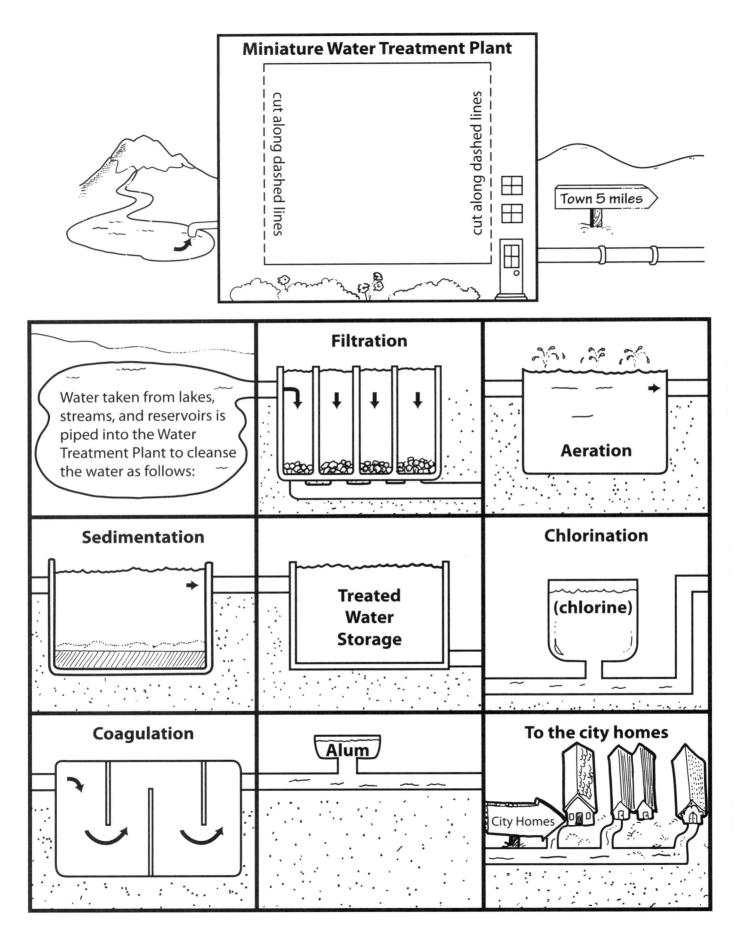

Miniature Water Treatment Plant

cut along dashed lines

cut along dashed lines

Town 5 miles

Water taken from lakes, streams, and reservoirs is piped into the Water Treatment Plant to cleanse the water as follows:

Filtration

Aeration

Sedimentation

Treated Water Storage

Chlorination

(chlorine)

Coagulation

Alum

To the city homes

City Homes

MINI Water Treatment Simulation

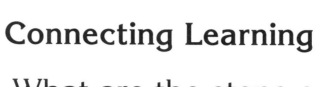

Connecting Learning

1. What are the steps of water treatment that we simulated?

2. What do you think would have happened if the steps were done in a different order?

3. How was our simulation like the real process of water treatment? How was it different?

4. Which part of the processes do you think had the greatest impact on cleaning the water?

MINi Water Treatment Simulation

Connecting Learning

5. How do the cleaned samples compare from group to group? What might be reasons for any differences?

6. Why aren't you allowed to drink the water in this simulation?

7. How could we find out how our local water is purified?

8. What are you wondering now?

Topic
Water quality

Key Question
How does the hardness of water affect the amount of suds the water will make?

Learning Goal
Students will observe, classify, and order water samples according to their degree of hardness and softness.

Guiding Documents
NRC Standards
- *Plan and conduct a simple investigation.*
- *Use mathematics in all aspects of scientific inquiry.*

NCTM Standards 2000*
- *Investigate how a change in one variable relates to a change in a second variable*
- *Collect data using observations, surveys, and experiments*
- *Understand such attributes as length, area, weight, volume, and size of angle and select the appropriate type of unit for measuring each attribute*
- *Select and apply appropriate standard units and tools to measure length, area, volume, weight, time, temperature, and the size of angles*

Math
Serial ordering
Measurement
    volume
    height

Science
Environmental science
    water quality
        hard and soft water

Integrated Processes
Observing
Predicting
Collecting and recording data
Comparing and contrasting
Controlling variables
Interpreting data
Generalizing

Materials
For the class:
    distilled water
    tap water
    bottled water
    salt water (see *Management 1*)
    food coloring

For each group:
    liquid dish soap
    four plastic water bottles with lids
      (see *Management 3)*
    graduated cylinder, 100 mL
    eyedropper
    metric rulers

For each student:
    Hard Water, Soft Water rubber band book
    #19 rubber band
    student page

Background Information
Pure water—that made up of only hydrogen and oxygen atoms—is very rarely found on Earth. Most water is a mixture of H_2O and many other minerals. Water collects minerals as it soaks into the soil and flows over rocks. Gradually, it becomes mineral rich.

Water that is full of minerals—generally calcium and magnesium ions—is called hard water. Soft water, by contrast, has very little calcium and magnesium relative to hard water. Another kind of water that is available is distilled water. Distilled water has had impurities removed through a process of boiling and collecting the steam in a clean container. The impurities are left behind when the water boils.

Hard water is problematic in homes for many reasons. Because of its high mineral content, it leaves deposits in the pipes that can cause them to become clogged. This is also true of fixtures such as faucets and showerheads. Hard water produces fewer suds and is not as good for cleaning.

For this reason, many people have soft water machines in their homes. These machines use salts to remove the calcium and magnesium from the tap water. This results in water that produces more suds and is better for cleaning purposes.

Management

1. To make the saltwater solution, mix one table-spoon of salt with one liter of tap water.
2. Use food coloring to color the water samples before giving them to students. Red: tap water; blue: distilled water; yellow: salt water; green: bottled water.
3. All the water bottles should be the same size and shape to facilitate comparisons between groups. Be sure to remove the labels.

Procedure

1. Ask students if they think all water will make the same amount of suds with equal amounts of soap and shaking.
2. Ask what variables might affect the amount of suds. If hard and soft water are not mentioned, add them to the list.
3. Distribute the rubber band book and read through it as a class. Explain that the softer the water, the more suds it will produce.
4. Tell the students they will do an experiment to test different waters for softness.
5. Have students get into groups of four and distribute the materials.
6. Discuss the color key on the student page and have students color the bottles to match.
7. Have students make and color their predictions about which water will produce the most suds.
8. Assign one type of water to each group member. Have each student fill his or her bottle with 100 mL of the water he or she has been assigned.
9. Instruct students to add one drop of liquid soap to each bottle and close the lid tightly.
10. Have the class shake their bottles for one minute. Stress the importance of students trying to use equal force while shaking their bottles.
11. After the minute is up, tell students to record the results and compare them to their predictions.
12. Have students measure the heights of the suds in their bottles as an objective measure of the difference between the types of water.

Connecting Learning

1. What is hard water? [water with dissolved minerals—especially calcium and magnesium] How does it become hard? [by coming in contact with Earth materials like soil and rocks]
2. How do soft water and distilled water compare to hard water? [Soft water has fewer dissolved minerals than hard water. Distilled water has had its impurities removed.]
3. Which kind of water did you predict would make the best bubbles? How did your predictions compare to the results?
4. Based on the results, which water sample is the softest? ...the hardest?
5. What generalization can you make about how water hardness relates to making suds? [the harder the water, the fewer the suds]
6. Which water would you choose to make bubbles? Why?
7. What are you wondering now?

Extensions

1. Have students use the *Hardly Soft* page to make hard and soft water and compare the amount of bubbles produced by each.
2. Make a survey and graph the results to show how many people have water softeners in their homes.
3. Test multiple brands of bottled water to see if they are hard or soft.

* Reprinted with permission from *Principles and Standards for School Mathematics*, 2000 by the National Council of Teachers of Mathematics. All rights reserved.

Key Question

How does the hardness of water affect the amount of suds the water will make?

Learning Goal

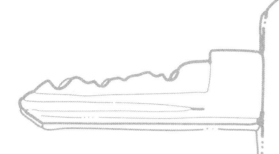

observe, classify, and order water samples according to their degree of hardness and softness.

You may have seen mineral deposits on your showerhead at home. If so, you probably have hard water. Hard water does not allow many soap suds to occur.

"My water's so hard, instead of a bubble bath I get a 'meteror shower!'"

Pure water is usually not found anywhere on Earth. Pure water would be only hydrogen and oxygen atoms. Most water is a mixture of H_2O and many other minerals.

Soft water produces more suds and is better for cleaning purposes. Many people have soft water machines in their homes. These machines use salts to remove the calcium and magnesium from the tap water.

water softener

HARD WATER, Soft Water

WATER, PRECIOUS WATER

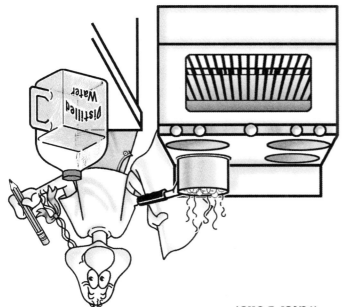

Distilled water has had impurities and minerals removed. It is made by boiling water and collecting the steam in a clean container. The impurities are left behind when the water boils.

Soft water has fewer dissolved minerals. Specifically, it has very little calcium and magnesium in it.

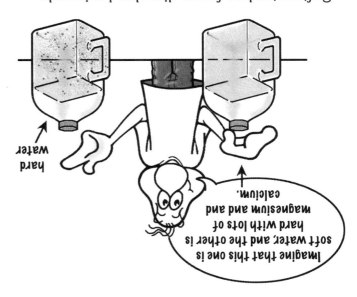

Imagine that this one is soft water, and the other is hard with lots of magnesium and and calcium.

hard water

calcium

Hard water causes problems in water pipes because it leaves minerals on the sides of the pipes. In time, these minerals can build up and clog the pipes.

Yucky mineral build-up

As rainwater soaks into the soil, it picks up minerals. Gradually it becomes mineral rich. Water that is full of minerals is called hard water.

6

3

Shake, Foam, and Suds

Does the hardness of water affect the amount of suds the water will make?

Color the key as shown.

A	B	C	D
Red	Blue	Yellow	Green
Tap Water	Distilled Water	Salt Water	Bottled Water

Make a prediction about which kind of water will make the most suds, and which will make the least. Use the colors from the key to color your prediction below.

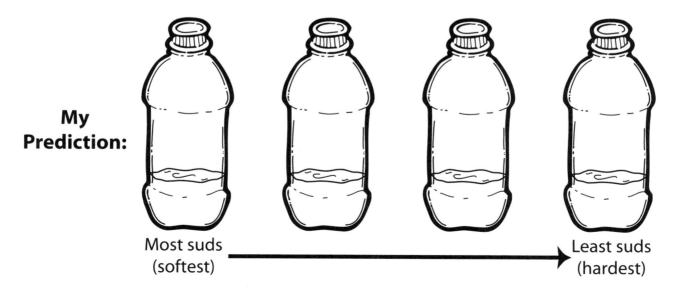

My Prediction:

Most suds (softest) → Least suds (hardest)

Shake, Foam, and Suds

You will now work with your group to test your predictions.
Each person in your group will test a different water sample.

1. Record your water type here: _____
2. Measure 100 mL of your water type and pour it into your bottle.
3. Add one drop of liquid soap to your bottle and close the lid tightly.
4. Shake the bottle for one minute.
5. Using a ruler, measure the height of the foam in each bottle.

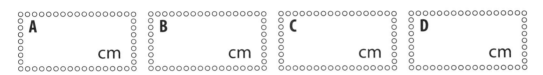

A	B	C	D
cm	cm	cm	cm

6. Record your actual results. Using the key, color the bottles from most suds to least suds.

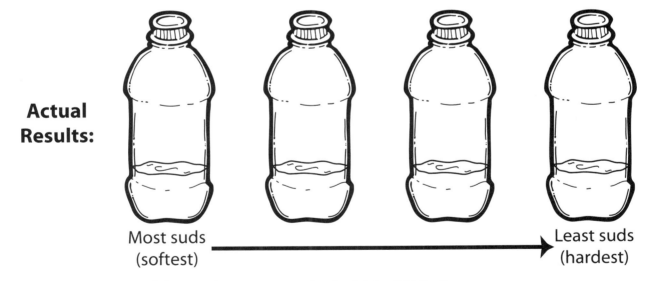

Actual Results:

Most suds   Least suds
(softest) ———————————————————————▶ (hardest)

7. Which water would you choose to make bubbles? Why?

8. Make a generalization about the kind of water that forms the best suds.

Hardly Soft

1. Fill two bottles with 100 mL of distilled water each. Label one bottle "Hard Water" and one bottle "Soft Water."

2. Add one teaspoon *Epsom salt* to the bottle labeled "Hard Water." Shake to dissolve the salt.

3. Add one teaspoon of *baking soda* to the bottle labeled "Soft Water." Shake to dissolve the baking soda.

4. Put two drops of soap in each bottle.

5. Shake the bottles for 30 seconds.

6. Measure and record the height of soap bubbles on top of the water.

7. Draw a picture showing the results.

8. Observe your bottles for 10 minutes. Describe your observations.

_____ cm
Soap bubble
height

_____ cm
Soap bubble
height

Hard Water   **Soft Water**

Observations:

Connecting Learning

1. What is hard water? How does it become hard?

2. How do soft water and distilled water compare to hard water?

3. Which kind of water did you predict would make the best bubbles? How did your predictions compare to the results?

4. Based on the results, which water sample is the softest? ...the hardest?

5. What generalization can you make about how water hardness relates to making suds?

6. Which water would you choose to make bubbles? Why?

7. What are you wondering now?

Red or Blue Will Tell You

Topic
Water quality/pollution

Key Question
What are some factors that can influence the quality of our drinking water?

Learning Goals
Students will:
- explore how contaminants can leach into drinking water, and
- model how water can be tested.

Guiding Documents
Project 2061 Benchmark
- *Fresh water, limited in supply, is essential for life and also for most industrial processes. Rivers, lakes, and groundwater can be depleted or polluted, becoming unavailable or unsuitable for life.*

NRC Standards
- *Human activities also can induce hazards through resource acquisition, urban growth, land-use decisions, and waste disposal. Such activities can accelerate many natural changes.*
- *Changes in environments can be natural or influenced by humans. Some changes are good, some are bad, and some are neither good nor bad. Pollution is a change in the environment that can influence the health, survival, or activities of organisms, including humans.*

NCTM Standard 2000*
- *Understand such attributes as length, area, weight, volume, and size of angle and select the appropriate type of unit for measuring each attribute*

Math
Measurement
  volume

Science
Enviromental science
  water pollution

Integrated Processes
Observing
Relating
Communicating
Collecting and recording data
Interpreting data
Inferring

Materials
For each group:
  small plastic tub (see *Management 1*)
  three coffee filters (see *Management 2*)
  soil (see *Management 3*)
  a wide permanent marker
  20 mL vinegar
  20 mL ammonia (see *Management 4*)
  three eyedroppers (see *Management 5*)
  litmus paper (see *Management 6*)
  three small portion cups
  plastic cup, 9 oz (see *Management 7*)

For the class:
  a water source

Background Information
Most of our country's rural residents use groundwater to supply their drinking water and farming needs. Wells are designed to provide clean water. If they are not properly maintained, they may allow bacteria, pesticides, fertilizer, or oil products to contaminate the groundwater. These contaminants can put our ecological system at risk. The proximity of a well to the source of contamination determines the risk it poses to the water in the well. A spill of pesticides being mixed and loaded near the well could result in the contamination of the water supply.

Preventing well water contamination is very important. Once the groundwater supplying a well is contaminated, it can be very difficult to clean up. The options may be to treat the water, drill a new well, or obtain water from another source. A contaminated well can also affect other nearby wells, posing a serious health threat to family and neighbors.

In this activity, litmus paper will be used to test the well water before and after a simulated fertilizer spill (ammonia) as well as a simulated acid rain runoff. Litmus is a substance extracted from organisms called lichens. It can be made in acidic form, which is red, or in basic form which is blue. If a liquid is acidic, it will turn blue litmus red, but it will not affect red litmus. If a liquid is basic, it will turn red litmus blue, but it will not affect blue litmus. A neutral liquid, which is neither acidic or basic, will not change the color of either form of litmus.

Management
1. Each group of four will need a small plastic tub. A plastic shoebox purchased at a local discount store works well. The surface area needs to be at least 35 cm square.

2. You may need to cut the coffee filters based on the depth of your container.
3. This activity was tested with a 50/50 potting soil to sand mixture. You may choose any soil mixture.
4. A window cleaner containing ammonia can be substituted for the ammonia. The window cleaner will have less of an odor and should not bother even an asthmatic student. If you are using the ammonia, dilute it 10:1 (water to ammonia) before giving it to your students. Follow the cautionary statement on the bottle when working with ammonia.
5. A drinking straw can be used for sampling the water if eyedroppers are not available.
6. To conserve litmus paper, have the students cut the strips in half. Litmus paper can be purchased at most local teacher stores or borrowed from a high school chemistry lab.
7. A measuring cup can be constructed from a nine-ounce clear plastic tumbler. Tape the calibrated strip to the side of the cup, aligning the zero with the bottom of the cup.
8. Although the soil used in this investigation will not hurt the ground water, model proper disposal techniques by collecting the soil used in the investigation in a large trash bag and properly disposing of it.

Procedure
1. Give each group of four a plastic tub. Instruct the students to fill the tub with the soil up to approximately five centimeters from the top of the container.
2. Instruct students to use the permanent marker, to label the coffee filters A, B, and C to identify the three wells.
3. Have the students construct the wells by placing a piece of coffee filter around the wide marker and inserting it into the soil. Remove the marker, leaving the filter in the soil. Repeat for all three wells.

4. Instruct them to add 100 mL of water to each well. Ask the groups to observe the wells as the water is added. Allow the well water to settle.
5. Have students label three portion cups and three eyedroppers, A, B, and C. Tell the students to take a small sample of the water from each well and place the samples in the corresponding cups.

Using both red and blue litmus paper, have the students test the water and attach the strips to the *Well Recording Sheet* in the appropriate place. Discuss the results.

6. Briefly discuss the fact that acid rain is a popular phrase used to describe precipitation containing sulfur that collects in the atmosphere due to the burning of fossil fuels such as coal, petroleum, and gasoline. Ask the students to measure out 30 mL of vinegar and tell them to simulate an acid rain runoff by pouring it in the soil to the left of well A. Allow a few minutes for it to seep into the ground.
7. Direct the students to empty and clean out the original three portion cups. Tell them to take a new sample from each of the wells and place it into the corresponding cup. Have the students repeat the litmus test and record just as they did earlier. Discuss the results.
8. Repeat the procedure using 30 mL of ammonia, pouring it in the soil to the right of well C. The ammonia will simulate a fertilizer runoff.
9. Compare the results from all three tests performed on the well water. Research possible solutions to contamination of well water.

Connecting Learning
1. How does a well work?
2. What would cause the water to go up or down in a well?
3. How did the contamination get into the well water?
4. How does soil help protect the groundwater? [Soil is a filtering agent for water.]
5. What types of things in your home, if poured on the ground, might contaminate drinking water?
6. Why do you think we have laws that tell us how to dispose of toxic household items?
7. What is acid rain and how does it affect our water supply?
8. What are you wondering now?

Extensions
1. Have the students test local water sources such as streams, lakes, aqueducts, etc.
2. Contact the local U.S. Geological Survey branch and gather data on the water in your area.
3. Further research the influence that contaminated well water will have on an aquifer.

* Reprinted with permission from *Principles and Standards for School Mathematics*, 2000 by the National Council of Teachers of Mathematics. All rights reserved.

Red or Blue Will Tell You

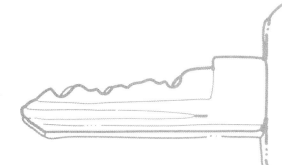

Key Question

What are some factors that can influence the quality of our drinking water?

Learning Goals

Students will:

- explore how contaminants can leach into drinking water, and

- model how water can be tested.

Red or Blue Will Tell You

Graduated Scales
for 9-oz cups

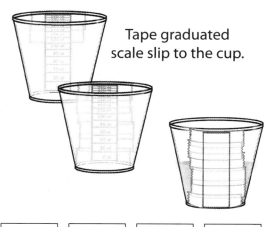

Tape graduated scale slip to the cup.

Cover completely with tape.

240	240	240	240	240	240	240	240	240	240
220	220	220	220	220	220	220	220	220	220
200	200	200	200	200	200	200	200	200	200
180	180	180	180	180	180	180	180	180	180
160	160	160	160	160	160	160	160	160	160
140	140	140	140	140	140	140	140	140	140
120	120	120	120	120	120	120	120	120	120
100	100	100	100	100	100	100	100	100	100
80	80	80	80	80	80	80	80	80	80
60	60	60	60	60	60	60	60	60	60
40	40	40	40	40	40	40	40	40	40
20	20	20	20	20	20	20	20	20	20
mL	mL	mL	mL	mL	mL	mL	mL	mL	mL

240	240	240	240	240	240	240	240	240	240
220	220	220	220	220	220	220	220	220	220
200	200	200	200	200	200	200	200	200	200
180	180	180	180	180	180	180	180	180	180
160	160	160	160	160	160	160	160	160	160
140	140	140	140	140	140	140	140	140	140
120	120	120	120	120	120	120	120	120	120
100	100	100	100	100	100	100	100	100	100
80	80	80	80	80	80	80	80	80	80
60	60	60	60	60	60	60	60	60	60
40	40	40	40	40	40	40	40	40	40
20	20	20	20	20	20	20	20	20	20
mL	mL	mL	mL	mL	mL	mL	mL	mL	mL

Red or Blue Will Tell You

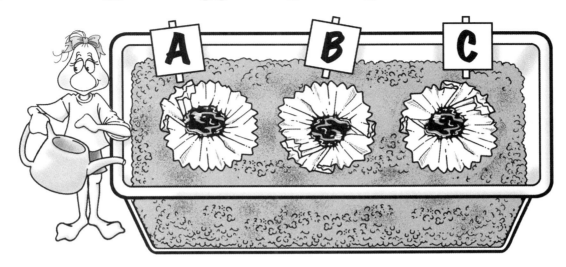

Tape or glue your litmus paper strips to each well below.

Beginning water samples.

A

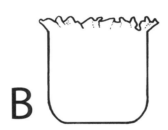
B

C

Water samples after an acid rain.

A

B

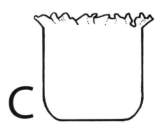
C

Water samples after a fertilizer spill.

A

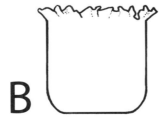
B

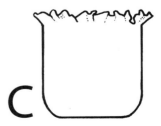
C

Explain your results on the back of this paper.

Red or Blue Will Tell You

Connecting Learning

1. How does a well work?

2. What would cause the water to go up or down in a well?

3. How did the contamination get into the well water?

4. How does soil help protect the groundwater?

5. What types of things in your home, if poured on the ground, might contaminate drinking water?

Red or Blue Will Tell You

Connecting Learning

CONNECTING LEARNING

6. Why do you think we have laws that tell us how to dispose of toxic household items?

7. What is acid rain and how does it affect our water supply?

8. What are you wondering now?

Little Sprouts

Topic
Water pollution

Key Question
What happens to your seeds when you water them with polluted water?

Learning Goal
Students will see the effects of watering seeds with polluted water.

Guiding Documents
Project 2061 Benchmark
- *Fresh water, limited in supply, is essential for life and also for most industrial processes. Rivers, lakes, and groundwater can be depleted or polluted, becoming unavailable or unsuitable for life.*

NRC Standards
- *Organisms have basic needs. For example, animals need air, water, and food; plants require air, water, nutrients, and light. Organisms can survive only in environments, and distinct environments support the life of different types of organisms.*
- *Changes in environments can be natural or influenced by humans. Some changes are good, some are bad, and some are neither good nor bad. Pollution is a change in the environment that can influence the health, survival, or activities of organisms, including humans.*

Science
Environmental science
    water pollution
Life science
    plants

Integrated Processes
Observing
Collecting and recording data
Comparing and contrasting
Generalizing

Materials
Part One:
    plastic sandwich bags
    paper towels
    pre-soaked lima beans
    watering solutions (see *Management 3*)
    transparent tape

Part Two:
    four sprouted lima bean plants (see *Management 5*)
    watering solutions (see *Management 3*)
    masking tape

Background Information
Water is essential for life. Most seeds will begin to germinate when soaked in water. Polluted water can have a detrimental effect on the plants in an area, causing them to have stunted growth or even to die. Polluted water can enter rivers, lakes, and streams as runoff after rains. Ground water and wells can be affected by runoff as well. In some parts of the developing world (e.g., China), water pollution is a huge problem that is severely impacting local populations of people, plants, and animals.

In some sections of the country, the salt content in the soil is very high and may affect the plant life in the area. The saltwater solution represents this condition. The soapy water represents pollutants from developed areas, such as detergents, lawn fertilizers, and household chemicals that can contaminate the water. The vinegar represents acids that could result from acid rain and other industrial chemicals that can find their way into our water supply. (Vinegar is about 5% acetic acid and 95% water.)

Management
1. This activity is divided into two parts. Each requires at least a week of observations. *Part One* has students look at the effect of polluted water on seed germination. *Part Two* examines the effects of pollution on growing plants. *Part Two* is optional and requires advance preparation. It can be done at the same time as *Part One*, or after *Part One* has been completed.
2. The day of the activity, pre-soak four large lima beans per student in a mixture of one part bleach to nine parts water. This will increase the speed of germination and help prevent seeds from molding.
3. You will need to make four watering solutions for the class. These solutions can be made and stored in two-liter soda bottles for easy access by the students. The solutions are as follows:
    solution one—tap water
    solution two—salt water—one cup salt per two-liter bottle
    solution three—soapy water—½ cup liquid soap per two-liter bottle
    solution four—vinegar—use distilled white vinegar right out of the bottle
4. Make a large bulletin board garden scene with four sections—one for each of the different solutions.
5. For *Part Two* of this activity, each group will need four sprouted lima bean plants. Plant the

lima beans in individual paper or plastic cups (8-10 oz). (Bagged lima beans from the grocery store work well.) You may want to plant a few extras in case there are some that don't sprout.

Procedure
Part One
1. Discuss what a seed needs in order to sprout. [sunlight, water, nutrients (usually from soil), air]
2. Ask students how water gets polluted. Discuss what they think the effects of polluted water would be on seeds. Have them record their thoughts on the first student page for *Part One.* Suggest that they try four different solutions and see what happens.
3. Divide the class into groups of four and assign a different solution to each student in the group. Explain what the different solutions represent (see *Background Information*).
4. Hand out one plastic bag, one paper towel, and four pre-soaked seeds to each student. Have students label their bags with their names and solutions.
5. Have the students fold the paper towels into fourths so that they will fit into the plastic bags.
6. Give each student a length of tape with which to adhere the beans to the paper towel. Have them position and tape the beans in the approximate center of the folded paper towel.

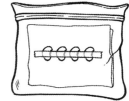

7. Tell them to dampen their paper towels with their watering solutions and then slide them into the plastic bags. Have them seal their bags.
8. Each bag may be stapled to a garden scene on the bulletin board or taped to a sunny window.
9. Have the students observe the seeds over a period of at least one week. Instruct them to record their observations and draw pictures of their seeds' growth on the student pages.
10. If necessary, have students add more watering solution to their plastic bags so that the paper towels stay damp.
11. At the end of the observation time, have students write their conclusions about the effects of pollution on seed growth. Discuss the results as a class and have groups compare their seeds to those of other groups.

Part Two (Optional)
1. Ask students if they think the effects of polluted water would be any different on plants that are growing than they would be on seeds that have not yet sprouted. Have them record their thoughts on the first student page for *Part Two.*

2. Give each student a lima bean plant growing in a cup. Have students use masking tape to label their cups with their names and the watering solution being used. (Students should use the same solutions that they did for *Part One.*)
3. Provide a sunny location for the plants where they will get the light needed each day. Have students water the plants as needed. Caution them not to over-water the plants.
4. Instruct students to record their observations on the student pages using words and pictures.
5. At the end of the observation time, have students write their conclusions about the effects of pollution on plant growth. Discuss the results as a class and have groups compare their plants to those of other groups.

Connecting Learning
Part One
1. Did all of the seeds start to grow?
2. Which seeds grew at the beginning of the experiment? ...at the end?
3. Why was it important to water one set of seeds with water? [to have a control group—a normal to use as a comparison]
4. What did the seeds look like when they stopped growing?
5. Why do you think they stopped growing?
6. What kind of pollution appeared to be the worst for the seeds?
7. Which liquid would you choose to water your seeds?

Part Two
1. How did your lima bean plant look at the beginning of the experiment?
2. What changes did you notice in your plant? How long did it take for these changes to appear?
3. How did the different plants in your group compare?
4. Which kind of pollution seemed to affect the plants the most?
5. How did the effects of pollution on the plants compare to the effects of pollution on the seeds?
6. How does the water in the real world become polluted?
7. How can we prevent water pollution?
8. What are you wondering now?

Extensions
1. Experiment with other types of polluted water following the same procedures.
2. See if there are other kinds of seeds or plants that appear to be resistant to some kinds of pollution. For example, there are some plants that have adapted to live in environments where there are high concentrations of salt in the water.

Little Sprouts

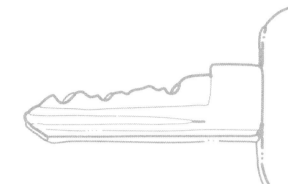

Key Question

What happens to your seeds when you water them with polluted water?

Learning Goal

Students will:

see the effects of watering seeds with polluted water.

Little Sprouts

Part One

How will polluted water affect seed growth?

I think:

I am watering my seeds with: _____ .

My observations: (use words and pictures)

Day _____

Day _____

Little Sprouts

Part One

My observations, continued:

Day _____

Day _____

Conclusions:

Little Sprouts

Part Two
How will polluted water affect plant growth?

I think:

I am watering my seeds with: _____

My observations: (use words and pictures)

Day _____

Day _____

Little Sprouts

Part Two

My observations, continued:

Day _____

Day _____

Conclusions:

Little Sprouts

Connecting Learning

Part One

1. Did all of the seeds start to grow?

2. Which seeds grew at the beginning of the experiment? ...at the end?

3. Why was it important to water one set of seeds with water?

4. What did the seeds look like when they stopped growing?

5. Why do you think they stopped growing?

6. What kind of pollution appeared to be the worst for the seeds?

7. Which liquid would you choose to water your seeds?

Little Sprouts

Connecting Learning

Part Two

1. How did your lima bean plant look at the beginning of the experiment?

2. What changes did you notice in your plant? How long did it take for these changes to appear?

3. How did the different plants in your group compare?

4. Which kind of pollution seemed to affect the plants the most?

5. How did the effects of pollution on the plants compare to the effects of pollution on the seeds?

Little Sprouts

Connecting Learning

6. How does the water in the real world become polluted?

7. How can we prevent water pollution?

8. What are you wondering now?

WATER FACTS

Topic
Water facts

Key Question
What do you know about water?

Learning Goal
Students will participate in a cooperative learning lesson in which they will be exposed to a variety of interesting facts about water.

Guiding Documents
Project 2061 Benchmarks
- *When liquid water disappears, it turns into a gas (vapor) in the air and can reappear as a liquid when cooled, or as a solid if cooled below the freezing point of water. Clouds and fog are made of tiny droplets of water.*
- *Fresh water, limited in supply, is essential for life and also for most industrial processes.*

NRC Standards
- *Materials can exist in different states—solid, liquid, and gas. Some common materials, such as water, can be changed from one state to another by heating or cooling.*
- *Water, which covers the majority of the earth's surface, circulates through the crust, oceans, and atmosphere in what is known as the "water cycle." Water evaporates from the earth's surface, rises and cools as it moves to higher elevations, condenses as rain or snow, and falls to the surface where it collects in lakes, oceans, soil, and in rocks underground.*

Science
Earth science
   water

Integrated Processes
Communicating
Recording data
Applying

Materials
For the class:
   32 question cards

Background Information
The background information for this activity is contained in previous activities as well as on the fact card solutions. Many of the facts will be things that students may not know the answer to, and that's okay These cards are designed to stimulate thinking and discussion; the logic used in discussing an answer is perhaps more important than the answer itself.

Management
1. Prepare the question cards ahead of time. They should be copied onto card stock and cut apart. They can be colored and laminated for future use.
2. Be sure the recorder is aware that each card has a number. The group's answer needs to be written on the correct number line.
3. Set up a system to pass the four cards on to the next group. Select a student from each group to pass the cards.
4. Caution students not to mark on the cards.

Procedure
1. Divide the class into groups of four students. Have each group select one person to record the group's responses and another to be in charge of switching cards with another group. The recorder needs a sheet of lined paper on which to record the group's responses.
2. Tell the class that they are going to be given 32 true/false statements. They are to discuss the questions and arrive at a group answer. The recorder will write down that answer. Emphasize that they may not know the correct answer for each question, but that they are to share ideas and come up with a group answer. Discuss the social skills necessary for reaching a group decision. Encourage all group members to participate. Discuss acceptable ways to disagree with someone's answer.
3. Give each group four of the question cards and tell them that they have a few minutes to finish the set. Be sure that they pay attention to the numbers on the cards they have and record their answers in the appropriate spaces.
4. Once a few minutes have passed, inform groups that their time is up, and have them pass their cards to the next group. Be sure to keep the rotation the same each time. If you have eight groups, each group will pass the cards eight times until all 32 questions have been answered.

249

5. Refer to the *Solutions* and read the answers and the explanatory statements. Discuss the answers with the students.

Connecting Learning
1. Which answers did you know going into the activity?
2. Which of the cards had answers that were surprising to you? Why?
3. What are some ways we can conserve water? Why is this important
4. What are you wondering now?

Extensions
1. Have students choose a card and make a water conservation poster based on the information on that card.
2. Challenge students to do some research on water and make up additional fact cards.
3. Devise activities that demonstrate some of the facts learned.

Solutions
These are the solutions to the cards. For answers with a star (*), accept responses from students—true or false—if the answer is based on some logical explanation.

1. Water will boil at 180° F.
   True*—at sea level water boils at 100°C (212°F), but the boiling point of water decreases as air pressure decreases. High in the mountains, water will boil at a lower temperature.
2. Steam is invisible.
   True—steam is invisible. What we usually think of as steam is actually condensed water.
3. Your body is 75% water.
   False—the amount of water in your body varies (usually between 45% and 75% of body weight) depending on the amount and type of body fat.
4. One oxygen atom and two hydrogen atoms can form a water molecule.
   True—H_2O—however, it is very rare to find a single water molecule.
5. Taking a five-minute shower uses about 18 gallons of water.
   False—about 12.5 gallons are used (2.5 gpm x 5 minutes = 12.5 gallons).
6. A person needs to take in 2 ½ quarts of water a day to stay healthy.
   True—but this includes water from all sources, including food.
7. Water is the only substance on earth that is naturally present in three forms—solid, liquid, and gas.
   True
8. 85% of the water on Earth is in the salty oceans.
   False—97% is in the salty oceans.

9. A gallon of water weighs just over 8 pounds.
   True—a gallon of water weighs 8.34 pounds.
10. Toilets account for about 40% of indoor water use in a home.
    False—toilets account for about 27% of residential indoor water use.
11. Water is the most common substance on Earth.
    True
12. It takes 115 gallons of water to grow enough wheat to make a loaf of bread.
    True
13. Every American uses 30 gallons of water a day in their homes.
    False—the average American uses about 60 gallons per day indoors and more than 170 gallons per day when leaks and outdoor use are taken into consideration.
14. Snow is a form of precipitation.
    True
15. Water will freeze at 32°C.
    False—water freezes at 32°F or 0°C.
16. Water will expand when heated.
    True*—it contracts from 0°C to 4°C, then expands.
17. Water is a good conductor of electricity.
    False*—pure water is not a good conductor. When a person is wet the person loses his/her resistance and becomes a good conductor.
18. Water contracts when frozen.
    False—it expands. It is rare for a substance to expand when frozen.
19. Animals cannot walk on liquid water.
    False—due to surface tension, many insects can.
20. Your brain is 75% water.
    True
21. While falling, a drop of water is shaped like a teardrop.
    False—due to surface tension, a free-falling drop is round.
22. Washing a car uses about 100 gallons of water.
    True*—a 10- to 15-minute car wash uses about 100 gallons if the hose is left running. (A typical garden hose uses 5-8 gallons of water per minute, but may use as much as 10 gpm.)
23. The prefix "hydro" means "water."
    True
24. A gallon of oil is heavier than a gallon of water.
    False—water is denser than oil, and an equal volume is heavier.
25. It takes 2000 gallons of water to manufacture a car.
    False—it takes about 39,000 gallons to manufacture a car.
26. Water is made up of two colorless and odorless gases.
    True—hydrogen and oxygen—H_2O.

27. Water can flow uphill.
    True—in capillary action in plants, water molecules are attracted to one another and are pulled up to the top of the plant.
28. The amount of water vapor in the air is called humidity.
    True
29. It is best to water the lawn when it is hot and windy.
    False—much of the water will evaporate and the water will not reach the plants' roots.
30. Adding salt to water will cause it to boil at a higher temperature.
    True—this is why you add salt to water when boiling pasta.
31. Dew is a form of condensation.
    True
32. A person could live without water for about a week.
    True*—however, this depends on the temperature of the environment.

WATER FACTS

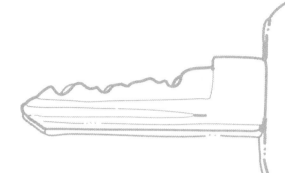

Key Question

What do you know about water?

Learning Goal

Students will:

participate in a cooperative learning lesson in which they will be exposed to a variety of interesting facts about water.

1. Water will boil at 180°F.
True or False

2. Steam is invisible.
True or False

3. Your body is 75% water.
True or False

4. hydrogen hydrogen oxygen
One oxygen atom and two hydrogen atoms can form a water molecule.
True or False

5. Taking a five-minute shower uses about 18 gallons of water.
True or False

6. A person needs to take in 2½ quarts of water a day to stay healthy.
True or False

7. Water is the only substance on Earth that is naturally present in three forms—solid, liquid, and gas.
True or False

8. 85% of the water on Earth is in the salty oceans.
True or False

9 A gallon of water weighs just over 8 pounds.

True or False

13 Every American uses 30 gallons of water a day in their home.

True or False

10 Toilets account for about 40% of indoor water use in a home.

True or False

14 Snow is a form of precipitation.

True or False

11 Water is the most common substance on Earth.

True or False

15 Water will freeze at 32°C.

True or False

12 It takes 115 gallons of water to grow enough wheat to make a loaf of bread.

True or False

16 Water will expand when heated.

True or False

17 Water is a good conductor of electricity.

True or False

18 Water contracts when frozen.

True or False

19 Animals cannot walk on liquid water.

True or False

20 Your brain is *75%* water.

True or False

21 While falling, a drop of water is shaped like a teardrop.

AHHH!

True or False

22 Washing a car uses about 100 gallons of water.

True or False

23 The prefix "hydro" means "water."

True or False

24 A gallon of oil is heavier than a gallon of water.

MOTOR OIL 20-50 wt.

True or False

25 2,000 GALLONS
It takes 2000 gallons of water to manufacture a car.
True or False

29 It is best to water the lawn when it is hot and windy.
True or False

26 Water is made up of two colorless and odorless gases.
Gas 1 + Gas 2 = Water
True or False

30 Adding salt to water will cause it to boil at a higher temperature.
True or False

27 Water can flow uphill.
True or False

31 Dew is a form or condensation.
True or False

28 The amount of water vapor in the air is called humidity.
True or False

32 A person could live without water for about a week.
True or False

WATER FACTS

Connecting Learning

1. Which answers did you know going into the activity?

2. Which of the cards had answers that were surprising to you? Why?

3. What are some ways we can conserve water? Why is this important

4. What are you wondering now?

Water Olympics

Topic
Properties of water

Key Question
What are some of the properties of water that we can observe in each experience?

Learning Goal
Students will participate in four events that help them to better understand the adhesive and cohesive forces involved with water.

Guiding Documents
NRC Standard
- *Employ simple equipment and tools to gather data and extend the senses.*

NCTM Standard 2000*
- *Select and apply appropriate standard units and tools to measure length, area, volume, weight, time, temperature, and the size of angles*

Math
Computation
Measurement
   length

Science
Physical science
   properties of water

Integrated Processes
Observing
Predicting
Collecting and recording data
Controlling variables

Materials
Amazing Water Race:
   roll of wax paper
   copies of water maze
   tape
   eyedropper
   toothpicks
   liquid dish soap
   timer or stopwatch

Fold and Float:
   aluminum foil cut in 12-cm squares
   bowls for water

Paper Towel Absorption:
   three different brands of paper towels
   metric rulers
   bowls or cups for water

Bubble Rings:
   liquid dish soap
   straws
   metric rulers
   cups

Background Information
Amazing Water Race/Water Stretch

Two forces are at work on the water molecules in this event—cohesion and adhesion. Water molecules are attracted to each other because of their molecular structure. This attraction of like molecules is called cohesion. This causes water molecules to want to stay together unless the cohesive bonds are weakened. Adhesion is the attraction between unlike molecules. Adhesion is what causes the water drops to "follow" the toothpick, allowing students to move the drops through the maze. The attraction of the water molecules to the molecules in the toothpick also allows the water drop to be stretched; the adhesive force is stronger, in this case, than the cohesive force.

Fold and Float

Aluminum should sink when placed in water because it has a density that is greater than that of water. However, when a piece of aluminum foil is placed flat on the surface of water, it will often float. This is because the surface tension of water is strong enough to hold up the aluminum foil even though it is 2.7 times denser than water. Surface tension is caused by the cohesion between water molecules. The molecules below the surface of the water are attracted equally in all directions, while those on the surface are only attracted to the sides and downward. This causes the surface of the water to contract and act like it is covered with a thin film. The surface tension of water is strong enough to hold up some objects that are more dense than water. This is why some insects, like the water strider, are able to walk on the surface of water.

Paper Towel Absorption

Water is able to travel through the narrow spaces between the fibers of paper towels by capillary action. The attractive force between the water molecules and the paper fibers is greater than the cohesive force between the water molecules. This causes the water molecules to be pulled up the paper towel against the force of gravity. The attraction between unlike molecules is called adhesion.

Bubble Rings

The surface tension of water is greater than that of other common liquids. Try blowing a bubble with water. It breaks apart because the surface tension (cohesive force) is so strong. Soap weakens the bonds between water molecules, reducing the surface tension to about one-third that of plain water. This makes the soapy water more elastic; it stretches to form bubbles when air is blown into it.

Management

1. These four activities may be done as individual lessons or as centers with students rotating through the activities. The task cards can be run off and placed at each center. Students should be responsible for cleaning up a center before moving on to the next one. An extra supply of paper towels may be placed at each center to facilitate clean up. It is important that these activities be followed by class discussions that focus on the water properties involved.

2. The procedures for each activity are given on the task cards, but some students may need each activity demonstrated before starting. If the students are doing the activities as part of a water Olympics, they will each need a copy of the score card. The students must make a prediction and record it on the score card before doing each event. The person with the lowest score is the winner. (If students have not had prior experience in exploring the adhesive and cohesive properties of water, their predictions will be merely guesses.)

Procedure

The following are special instructions for the four events.

Amazing Water Race/Water Stretch

Tape a piece of wax paper over each maze before starting. After doing the two activities for this event, you may want to have students observe the effect of soap on the cohesion of water by dipping a toothpick into liquid soap and then touching a large water drop with it. Make sure that you have fresh water, toothpicks, and wax paper if you repeat this activity or the soap left on the maze or toothpick will spoil the results for the next group.

Fold and Float

This activity could be extended to cover fractions. Each time you fold the foil in half you are decreasing its area by a power of two. After three folds you have only one-eighth of the original surface area, after four folds you have only one-sixteenth.

Paper Towel Absorption

The paper towels can be cut beforehand into strips. The school's paper towel can be used as one of the three brands tested for absorption rate. The students can tape the three strips to a pencil

so they can dip them simultaneously into the bowl of water.

Bubble Rings

Mix the bubble solution beforehand by adding 30 mL (two tablespoons) of liquid dish soap to the water in a two liter plastic bottle. Place four to six cups of bubble solution on the table along with a box of straws. The students will get their own straws when blowing bubbles and then will use centimeter rulers to measure the diameter of the ring that is left on the table when the bubble bursts.

Connecting Learning

1. Why were you able to pull the water drop through the maze? [Two forces were at work. The adhesive force of the water's attraction to the toothpick caused it to "follow" the toothpick. The cohesive force of the molecules within the drop to each other caused the drop to stay together.]

2. What did you have to do if the water drop broke into smaller drops? [Go back and pick them up.] Why were you able to do this? [Because the water was attracted to the water (cohesion) and would reform into a bigger drop.]

3. How were you able to stretch the water drop? What force was being displayed? [We used two toothpicks and stretched the drop as far as possible. This was evidence of adhesion because the water molecules were attracted to the toothpick's molecules.]

4. Was anyone able to stretch the water the entire length of the measuring strip? If so, how was it done? (Some students may discover that by laying the toothpicks along the measuring strip, the water drop can go the entire length.)

5. Why did everyone need to start with drops that filled the entire circle at the beginning of both of these experiences? [to make for a fair test and for fair comparisons]

6. Who was able to get the most folds of their aluminum foil boat? How many folds was that?

7. Why don't you think we could get smaller surface areas to float? [They became too dense to be held up by the surface tension of the water.]

8. What force causes the water to have surface tension? [cohesion]

9. Why can you blow bubbles with soapy water and you can't with plain water? [The cohesive force of plain water is too strong. Soap reduces the cohesive force of the water, making it more elastic.]

10. What are you wondering now?

*   Reprinted with permission from *Principles and Standards for School Mathematics*, 2000 by the National Council of Teachers of Mathematics. All rights reserved.

WATER OLYMPICS

Key Question
What are some of the properties of water that we can observe in each experience?

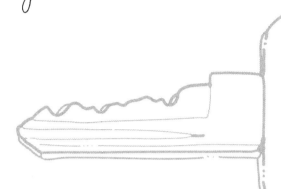

Learning Goal

Students will:

participate in four events that help them to better understand the adhesive and cohesive forces involved with water.

WATER OLYMPICS

Event	Prediction	Actual	Difference*
Amazing Water Race	_____ seconds	_____ seconds	
Water Stretch	_____ cm	_____ cm	
Fold 'n' Float	_____ folds	_____ folds	
Paper Towel Absorption	Paper towel # _____	Paper towel # _____	Right = 0 Wrong = 5 _____
Bubble Rings	_____ cm	_____ cm	
		Total Differences	

* Subtract the lower number from the higher number.

AMAZING WATER RACE

Tape a piece of wax paper on top of this page.

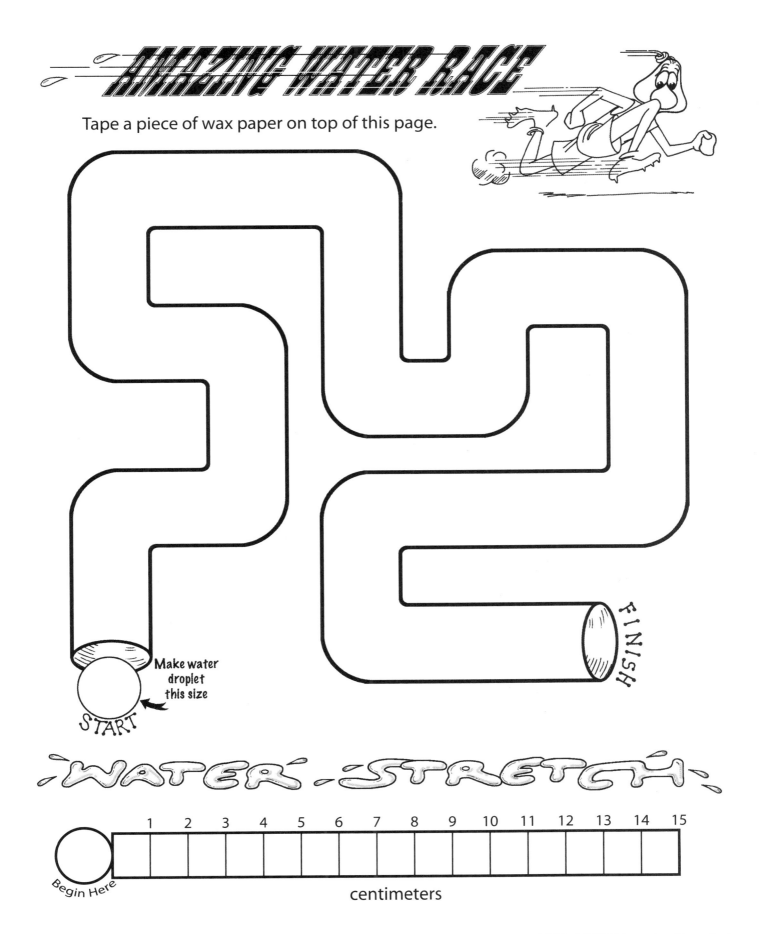

Make water droplet this size

START

FINISH

WATER STRETCH

	1	2	3	4	5	6	7	8	9	10	11	12	13	14	15

Begin Here

centimeters

How can your water drop be guided through the maze?

Procedure:
1. Tape a piece of wax paper on top of the maze.
2. Place a water drop to fit inside the circle on your paper.
3. Guess how many seconds it will take to move the water drop through the maze. Record your guess in the table.
4. Move the water drop through the maze with a toothpick. If the drop separates, go back and collect it before you continue.
5. Record how long it actually took you to move through the maze. Compare this time to your guess.

How long can you stretch a drop of water to be?

Procedure:
1. Guess how far can you stretch the drop of water. Record your guess in the table.
2. Make a drop of water the size of the circle.
3. Use toothpicks to stretch the water drop.
4. Record how long you were able to stretch it.
5. Find the difference between your guess and the actual length.

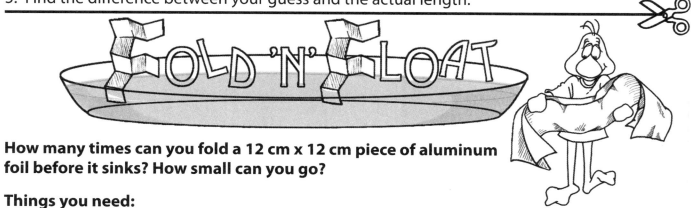

How many times can you fold a 12 cm x 12 cm piece of aluminum foil before it sinks? How small can you go?

Things you need:
- one bowl (two-thirds full of water)
- one 12 cm x 12 cm piece of aluminum foil

Procedure:
1. Guess the number of folds that will cause the foil to sink. Record your guess.
2. Float the piece of foil in the water.
3. Fold it in half—that's your first fold—place it back in the water. Does it float?

4. If so, fold it in half again—that's your second fold. Does it float?
5. Keep folding the foil in half and testing whether or not it floats after each fold.
6. Keep making the surface area of the foil smaller until it sinks.

PAPER TOWEL Absorption

Which brand of paper towels absorbs the water the fastest?

You need:
- one 3 cm x 20 cm strip of each brand of paper towel per group
- one bowl of water
- metric ruler

Procedure:
1. Guess which brand will absorb water the fastest. Record your guess.
2. Mark each strip at the 18 cm mark.
3. Place one strip from each of the test strips into a bowl of water all at the same time.
4. Record which strip the water reaches the 18 cm mark on first.
5. Compare the guess with the actual results. If your guess was correct, give yourself a difference of zero. If not, give yourself five points in the difference column.

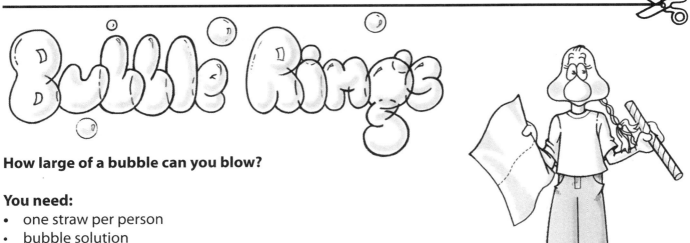

How large of a bubble can you blow?

You need:
- one straw per person
- bubble solution
- metric ruler

Procedure:
1. Guess how large of a bubble you will be able to blow. Record your guess.
2. Wet the table top with bubble solution.
3. Trap some bubble solution in the straw.
4. Place the straw on the table top at an angle and blow gently.
5. Measure and record the diameter of bubble after it pops.

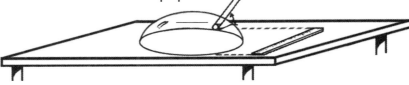

WATER OLYMPICS

Connecting Learning

1. Why were you able to pull the water drop through the maze?

2. What did you have to do if the water drop broke into smaller drops? Why were you able to do this?

3. How were you able to stretch the water drop? What force was being displayed?

4. Was anyone able to stretch the water the entire length of the measuring strip? If so, how was it done?

Connecting Learning

5. Why did everyone need to start with drops that filled the entire circle at the beginning of both of these experiences?

6. Who was able to get the most folds of their aluminum foil boat? How many folds was that?

7. Why don't you think we could get smaller surface areas to float?

8. What force causes the water to have surface tension?

9. Why can you blow bubbles with soapy water and you can't with plain water?

10. What are you wondering now?

The AIMS Program

AIMS is the acronym for "Activities Integrating Mathematics and Science." Such integration enriches learning and makes it meaningful and holistic. AIMS began as a project of Fresno Pacific University to integrate the study of mathematics and science in grades K-9, but has since expanded to include language arts, social studies, and other disciplines.

AIMS is a continuing program of the non-profit AIMS Education Foundation. It had its inception in a National Science Foundation funded program whose purpose was to explore the effectiveness of integrating mathematics and science. The project directors, in cooperation with 80 elementary classroom teachers, devoted two years to a thorough field-testing of the results and implications of integration.

The approach met with such positive results that the decision was made to launch a program to create instructional materials incorporating this concept. Despite the fact that thoughtful educators have long recommended an integrative approach, very little appropriate material was available in 1981 when the project began. A series of writing projects ensued, and today the AIMS Education Foundation is committed to continuing the creation of new integrated activities on a permanent basis.

The AIMS program is funded through the sale of books, products, and professional-development workshops, and through proceeds from the Foundation's endowment. All net income from programs and products flows into a trust fund administered by the AIMS Education Foundation. Use of these funds is restricted to support of research, development, and publication of new materials. Writers donate all their rights to the Foundation to support its ongoing program. No royalties are paid to the writers.

The rationale for integration lies in the fact that science, mathematics, language arts, social studies, etc., are integrally interwoven in the real world, from which it follows that they should be similarly treated in the classroom where students are being prepared to live in that world. Teachers who use the AIMS program give enthusiastic endorsement to the effectiveness of this approach.

Science encompasses the art of questioning, investigating, hypothesizing, discovering, and communicating. Mathematics is a language that provides clarity, objectivity, and understanding. The language arts provide us with powerful tools of communication. Many of the major contemporary societal issues stem from advancements in science and must be studied in the context of the social sciences. Therefore, it is timely that all of us take seriously a more holistic method of educating our students. This goal motivates all who are associated with the AIMS Program. We invite you to join us in this effort.

Meaningful integration of knowledge is a major recommendation coming from the nation's professional science and mathematics associations. The American Association for the Advancement of Science in *Science for All Americans* strongly recommends the integration of mathematics, science, and technology. The National Council of Teachers of Mathematics places strong emphasis on applications of mathematics found in science investigations. AIMS is fully aligned with these recommendations.

Extensive field testing of AIMS investigations confirms these beneficial results:
1. Mathematics becomes more meaningful, hence more useful, when it is applied to situations that interest students.
2. The extent to which science is studied and understood is increased when mathematics and science are integrated.
3. There is improved quality of learning and retention, supporting the thesis that learning which is meaningful and relevant is more effective.
4. Motivation and involvement are increased dramatically as students investigate real-world situations and participate actively in the process.

We invite you to become part of this classroom teacher movement by using an integrated approach to learning and sharing any suggestions you may have. The AIMS Program welcomes you!

AIMS Education Foundation Programs

When you host an AIMS workshop for elementary and middle school educators, you will know your teachers are receiving effective, usable training they can apply in their classrooms immediately.

AIMS Workshops are Designed for Teachers
- Correlated to your state standards;
- Address key topic areas, including math content, science content, and process skills;
- Provide practice of activity-based teaching;
- Address classroom management issues and higher-order thinking skills;
- Give you AIMS resources; and
- Offer optional college (graduate-level) credits for many courses.

AIMS Workshops Fit District/Administrative Needs
- Flexible scheduling and grade-span options;
- Customized (one-, two-, or three-day) workshops meet specific schedule, topic, state standards, and grade-span needs;
- Prepackaged four-day workshops for in-depth math and science training available (includes all materials and expenses);
- Sustained staff development is available for which workshops can be scheduled throughout the school year;
- Eligible for funding under the Title I and Title II sections of No Child Left Behind; and
- Affordable professional development—consecutive-day workshops offer considerable savings.

University Credit—Correspondence Courses
AIMS offers correspondence courses through a partnership with Fresno Pacific University.
- Convenient distance-learning courses—you study at your own pace and schedule. No computer or Internet access required!

Introducing AIMS State-Specific Science Curriculum
Developed to meet 100% of your state's standards, AIMS' State-Specific Science Curriculum gives students the opportunity to build content knowledge, thinking skills, and fundamental science processes.
- Each grade-specific module has been developed to extend the AIMS approach to full-year science programs. Modules can be used as a complete curriculum or as a supplement to existing materials.
- Each standards-based module includes mathreading, hands-on investigations, and assessments.

Like all AIMS resources, these modules are able to serve students at all stages of readiness, making these a great value across the grades served in your school.

For current information regarding the programs described above, please complete the following form and mail it to: P.O. Box 8120, Fresno, CA 93747.

Information Request

Please send current information on the items checked:

____ *Basic Information Packet* on AIMS materials

____ Hosting information for AIMS workshops

____ AIMS State-Specific Science Curriculum

Name: _____

Phone:_____ E-mail:_____

Address: _____
         Street   City   State   Zip

AIMS Magazine

YOUR K-9 MATH AND SCIENCE CLASSROOM ACTIVITIES RESOURCE

The AIMS Magazine is your source for standards-based, hands-on math and science investigations. Each issue is filled with teacher-friendly, ready-to-use activities that engage students in meaningful learning.

- *Four issues each year (fall, winter, spring, and summer).*

Current issue is shipped with all past issues within that volume.

1824	Volume XXIV	2009-2010	$19.95
1825	Volume XXV	2010-2011	$19.95

Two-Volume Combination

M20810	Volumes XXIII & XXIV	2008-2010	$34.95
M20911	Volumes XXIV & XXV	2009-2011	$34.95

Complete volumes available for purchase:

1802	Volume II	1987-1988	$19.95
1804	Volume IV	1989-1990	$19.95
1805	Volume V	1990-1991	$19.95
1807	Volume VII	1992-1993	$19.95
1808	Volume VIII	1993-1994	$19.95
1809	Volume IX	1994-1995	$19.95
1810	Volume X	1995-1996	$19.95
1811	Volume XI	1996-1997	$19.95
1812	Volume XII	1997-1998	$19.95
1813	Volume XIII	1998-1999	$19.95
1814	Volume XIV	1999-2000	$19.95
1815	Volume XV	2000-2001	$19.95
1816	Volume XVI	2001-2002	$19.95
1818	Volume XVIII	2003-2004	$19.95
1819	Volume XIX	2004-2005	$19.95
1820	Volume XX	2005-2006	$19.95
1821	Volume XXI	2006-2007	$19.95
1822	Volume XXII	2007-2008	$19.95
1823	Volume XXIII	2008-2009	$19.95

Volumes II to XIX include 10 issues.

Call 1.888.733.2467 or go to www.aimsedu.org

Subscribe to the AIMS Magazine

$19.95 a year!

AIMS Magazine is published four times a year.

Subscriptions ordered at any time will receive all the issues for that year.

AIMS Online—www.aimsedu.org

To see all that AIMS has to offer, check us out on the Internet at www.aimsedu.org. At our website you can search our activities database; preview and purchase individual AIMS activities; learn about state-specific science, college courses, and workshops; buy manipulatives and other classroom resources; and download free resources including articles, puzzles, and sample AIMS activities.

AIMS News

While visiting the AIMS website, sign up for AIMS News, our FREE e-mail newsletter.
Included in each month's issue you will find:

- Information on what's new at AIMS (publications, materials, state-specific science modules, etc.)
- A special money-saving offer for a book and/or product; and
- Free sample activities.

Sign up today!

AIMS Program Publications

Actions with Fractions, 4-9
The Amazing Circle, 4-9
Awesome Addition and Super Subtraction, 2-3
Bats Incredible! 2-4
Brick Layers II, 4-9
The Budding Botanist, 3-6
Chemistry Matters, 4-7
Counting on Coins, K-2
Cycles of Knowing and Growing, 1-3
Crazy about Cotton, 3-7
Critters, 2-5
Earth Book, 6-9
Electrical Connections, 4-9
Exploring Environments, K-6
Fabulous Fractions, 3-6
Fall into Math and Science*, K-1
Field Detectives, 3-6
Finding Your Bearings, 4-9
Floaters and Sinkers, 5-9
From Head to Toe, 5-9
Fun with Foods, 5-9
Glide into Winter with Math and Science*, K-1
Gravity Rules! 5-12
Hardhatting in a Geo-World, 3-5
Historical Connections in Mathematics, Vol. I, 5-9
Historical Connections in Mathematics, Vol. II, 5-9
Historical Connections in Mathematics, Vol. III, 5-9
It's About Time, K-2
It Must Be A Bird, Pre-K-2
Jaw Breakers and Heart Thumpers, 3-5
Looking at Geometry, 6-9
Looking at Lines, 6-9
Machine Shop, 5-9
Magnificent Microworld Adventures, 5-9
Marvelous Multiplication and Dazzling Division, 4-5
Math + Science, A Solution, 5-9
Mathematicians are People, Too
Mathematicians are People, Too, Vol. II
Mostly Magnets*, 3-6
Movie Math Mania, 6-9
Multiplication the Algebra Way, 6-8
Off the Wall Science, 3-9
Out of This World, 4-8
Paper Square Geometry:
    The Mathematics of Origami, 5-12

Puzzle Play, 4-8
Pieces and Patterns*, 5-9
Popping With Power, 3-5
Positive vs. Negative, 6-9
Primarily Bears*, K-6
Primarily Earth, K-3
Primarily Magnets, K-2
Primarily Physics*, K-3
Primarily Plants, K-3
Primarily Weather, K-3
Problem Solving: Just for the Fun of It! 4-9
Problem Solving: Just for the Fun of It! Book Two, 4-9
Proportional Reasoning, 6-9
Ray's Reflections, 4-8
Sensational Springtime, K-2
Sense-Able Science*, K-1
Shapes, Solids, and More: Concepts in Geometry, 2-3
The Sky's the Limit, 5-9
Soap Films and Bubbles, 4-9
Solve It! K-1: Problem-Solving Strategies, K-1
Solve It! 2nd: Problem-Solving Strategies, 2
Solve It! 3rd: Problem-Solving Strategies, 3
Solve It! 4th: Problem-Solving Strategies, 4
Solve It! 5th: Problem-Solving Strategies, 5
Solving Equations: A Conceptual Approach, 6-9
Spatial Visualization, 4-9
Spills and Ripples, 5-12
Spring into Math and Science*, K-1
Statistics and Probability, 6-9
Through the Eyes of the Explorers, 5-9
Under Construction, K-2
Water Precious Water, 2-6
Weather Sense: Temperature, Air Pressure, and Wind, 4-5
Weather Sense: Moisture, 4-5
What's Next, Volume 1, 4-12
What's Next, Volume 2, 4-12
What's Next, Volume 3, 4-12
Winter Wonders, K-2

Essential Math
Area Formulas for Parallelograms, Triangles, and Trapazoids, 6-8
Circumference and Area of Circles, 5-7
The Pythagorean Relationship, 6-8

Spanish Edition
Constructores II: Ingeniería Creativa Con Construcciones
    LEGO®, 4-9
    The entire book is written in Spanish. English pages not included.

* Spanish supplements are available for these books. They are only
  available as downloads from the AIMS website. The supplements
  contain only the student pages in Spanish; you will need the English
  version of the book for the teacher's text.

For further information, contact:
AIMS Education Foundation • P.O. Box 8120 • Fresno, California 93747-8120
www.aimsedu.org • 559.255.6396 (fax) • 888.733.2467 (toll free)

Duplication Rights

No part of any AIMS books, magazines, activities, or content—digital or otherwise—may be reproduced or transmitted in any form or by any means—including photocopying, taping, or information storage/retrieval systems—except as noted below.